Praise for *GMO Myths and Truths*

"Many who defend the use of GMO crops and foods claim that there is no evidence that any GMO is harmful to health or the environment. But this is wrong. There is plenty of sound empirical evidence of such harm, presented by qualified scientists in peer-reviewed literature. This book is a succinct summary and documentation of that evidence."

– **Richard Jennings, PhD,** Department of History and Philosophy of Science, University of Cambridge, UK

"*GMO Myths and Truths* is a very important contribution to the GMO debate. Written in a clear and concise style, it is an invaluable reference for those wanting to learn more about the arguments of proponents and those critical of the technology. I found it of great value in my own efforts to better understand the issue. I hope this information will be widely available to students and to the general public."

– **Jane Goodall, PhD, DBE,** and UN Messenger of Peace

"Using peer-reviewed studies and other documented evidence, *GMO Myths and Truths* deconstructs the false and misleading claims that are frequently made about the safety and efficacy of GM crops and foods. The book shows that far from being necessary to feed the world, GM crops are a risky distraction from the real causes of hunger. What is more, there is no reason to take this risk, since GM crops do not consistently raise yields, reduce pesticide use, or provide more nutritious food. GM crops and foods have not been shown to be safe to eat – and both animal feeding studies and non-animal laboratory experiments indicate that some GM foods, as well as most of the chemicals required to produce them, are toxic. Fortunately, the book shows that there are effective and sustainable alternatives to GM that can ensure a safe and plentiful food supply for current and future populations. *GMO Myths and Truths* is an invaluable and easy-to-read resource [illegible]one, including students, scientists, and members of the general public."

– **David Schubert, PhD,** Professor [illegible] Cellular Neurobiology, Salk Institute for Biological [illegible] California, USA

"Genetic engineering and its applications in agriculture are complex topics, even for scientists that are engaged with them. The range of techniques is large, as is the range of their implications for different groups in society, from the farmer, through to the citizen, to the corporation. It is difficult sometimes to have the latest information at the end of the fingertips and to understand the difference in how the same evidence is framed depending on the interests behind the science. *GMO Myths and Truths* is a great resource for me. It is written in a form that makes it readable to non-experts without losing value for the expert looking for the latest references. It has become one of my standard 'go-to' reviews when I need to refresh myself on the history, practice, or technical details in this fast moving, demanding and important area."

– **Jack Heinemann, PhD,** Professor, School of Biological Sciences, University of Canterbury, New Zealand

"*GMO Myths and Truths* is a much needed objective evaluation to separate propaganda and failed promises from the science of this critical topic."

– **Don M. Huber, PhD,** Professor Emeritus, Purdue University

"The role of provider of children's health care has become an onerous job in this modern era. The rapid demise of children's health where more children are chronically sick than well is statistically demonstrated in the US. The common denominator to all children's health is their food. We, as pediatricians, have had to become geneticists and toxicologists to understand the impact of our changing food system in our modern agro-environment and its effect on our children's well-being without support from our traditional medical literary sources. Drs Antoniou and Fagan and Claire Robinson have provided a concise reference on the topic of GM foods and their impact that will educate the physician, farmer, citizen and policymaker. From this easily-readable third edition, a busy practitioner is able to extract the necessary information on the state of GM science to be utilized in their medical practice. Thoroughly sourced, this book is a must-read to understand the effects of GM food, Bt toxin, and pesticides, including their impacts on health. Its value is undeniable. It has simplified a very complex science without losing its scientific profundity."

– **Michelle Perro, MD,** Institute for Health and Healing, Greenbrae, California, USA

"To all those concerned citizens, policymakers, journalists, researchers and students who have missed a reliable, unbiased, comprehensive and readable account of plant genetic engineering and its unpredictable consequences: Here it is, and it is a splendid piece of work!"

– **Terje Traavik, PhD,** Professor of Virology and Professor Emeritus of Gene Ecology, Faculty of Health Sciences, UiT – The Arctic University of Norway; and Scientific Director Emeritus, Genøk – Centre for Biosafety, Norway

"The global GM commercial and political juggernaut has already corrupted science by promoting a propaganda campaign to misinform the world that 'science' has revealed the unqualified safety of GM crops and foods, with no further question. This vitally important book documents the wide-ranging and well-grounded science that demolishes this political deception – a deception that has traduced the good name of science. The book also shows the systematic double standards exercised by those promoting GM, whereby studies claiming to show GM is safe are accepted with lower standards of scientific rigour than those showing evidence of harm and the need for further research. This book is essential reading for all involved, anywhere, with GM and agrichemicals risk and policy issues."

– **Brian Wynne, PhD,** Professor Emeritus of Science Studies and founding Research Director of the Centre for the Study of Environmental Change (CSEC); and former Associate Director of the UK ESRC Centre for Economic and Social Aspects of Genomics, Cesagen, Lancaster University, UK

"*GMO Myths and Truths* is unquestionably the go-to source for up-to-date, scientifically rigorous evidence bearing on genetic modification. As a scientist, I applaud not simply the heavy reliance on refereed literature, but also the attention to organization and interpretation, which makes technical information accessible to the non-specialist."

– **E. Ann Clark, PhD,** Associate Professor (retired), Plant Agriculture, University of Guelph, Canada

"*GMO Myths and Truths* is a scientifically correct, honest, and accessible discussion of molecular biotechnology that demonstrates why corporate claims about the inherent safety of GMOs and their centrality in reducing world hunger are utter nonsense."

– **Sheldon Krimsky, PhD,** Professor, Tufts University, USA; co-editor, *The GMO Deception*

"The condensed and updated *GMO Myths and Truths* is a thoroughly researched, evidence-based synthesis of the latest information regarding the environmental and health consequences of genetically-modified crops. Updated and adapted from the well-received *GMO Myths and Truths*, it refutes common myths surrounding the alleged safety and effectiveness of GMOs and explains, in highly readable fashion, the scientific truths regarding this risky technology. I have found it valuable in my teaching and policy work, and highly recommend it for scientists, policymakers, educators, students, and indeed all individuals concerned about the corporate commodification of the world's food supply."

- **Martin Donohoe, MD, FACP,** Adjunct Associate Professor, School of Community Health, Portland State University; Member, Social Justice Committee, Physicians for Social Responsibility; Member, Board of Advisors, Oregon Physicians for Social Responsibility; Senior Physician, Internal Medicine, Kaiser Sunnyside Medical Center, USA

"This is an excellent and urgently needed scientific critique of GM crops."

- **Carlo Leifert,** Professor for Ecological Agriculture, Newcastle University, Newcastle-upon-Tyne, UK

"*GMO Myths and Truths* provides an unprecedented overview of the current situation regarding genetically modified food production and consumption. Topics covered include the genetic engineering process, the safety questions around GM foods, and agronomic and environmental problems during GM crop cultivation, including weed and pest resistance and the consequences of contamination. This book will prove extremely useful for scientists, policymakers and students. In addition, it will inform members of the public, activists, farmers, and consumers."

- **Mohamed Habib, PhD**, Full Professor, UNICAMP, Brazil

"*GMO Myths and Truths* is a comprehensive study of the myths and facts surrounding genetically modified foods and crops. It illuminates aspects of the topic that are commonly misunderstood. The issues are systematically explained and analyzed in simple language, which is nonetheless precise. I would recommend this book to postgraduate students of genetics as well as interested members of the public. I am confident that it will prove a success worldwide and within Brazil."

- **Nagib Nassar, PhD** (Genetics); Professor Emeritus, University of Brasília, Brazil

"An opt-out clause used by the members of many governments is that they 'will make no decisions unless backed by sound scientific evidence'. This blind faith in the honesty and integrity of all scientists has proved, repeatedly, to be based on false assumptions. Politicians often fail to recognize that scientists and manufacturers are, like the rest of us, human. They can fall prey to the same ambitions, to financial greed, to a hunger for prestige, and to manipulating their findings in order to satisfy these cravings, just like some in other walks of life. Politicians like black-and-white – there can be no shades of grey or room for doubt, and they leave little scope for scientists to admit that they don't know, so those with integrity are caught in a cleft stick. Do they speak out and risk being pilloried by those in power, or do they quietly keep their heads down?

"Nowhere is the controversy greater than in the scientific research upon which the reputed safety of genetically modified organisms is founded. We should be extremely grateful to the authors of *GMO Myths and Truths* for having the courage to review the research and to tell the world of their findings."

– **Margaret, Countess of Mar,** independent cross-bench member of the House of Lords of the UK Parliament and an elected hereditary peer

GMO MYTHS AND TRUTHS

CONDENSED AND UPDATED

A citizen's guide to the evidence on the safety and efficacy of genetically modified crops and foods

Claire Robinson, MPhil
Michael Antoniou, PhD
John Fagan, PhD

Third edition, Version 1.0

Distributed in the U.S. by Chelsea Green Publishing

GMO Myths and Truths (condensed and updated third edition)
Version 1.0
A citizen's guide to the evidence on the safety and efficacy of genetically modified crops and foods

Published by Earth Open Source. Third edition, Version 1.0

ISBN: 978-0-9934367-0-3

Earth Open Source
2nd Floor, 145–157 St John Street
London EC1V 4PY
Great Britain
www.earthopensource.org

Design: Peter Brown
Printed in Great Britain
Distributed in the U.S. by Chelsea Green Publishing

First edition of *GMO Myths and Truths* published in Great Britain in 2012 by Earth Open Source
Second edition of *GMO Myths and Truths* published in Great Britain in 2014 by Earth Open Source

Disclaimer

The views and opinions expressed in this report are those of the individual authors and do not represent the official policy, position, or views of any organizations or institutions that the authors may be affiliated with.

Acknowledgements

The authors would like to thank the many scientists and experts who have contributed to the *GMO Myths and Truths* project from its inception in 2010, as well as all those who have read the material and put it to use.

Dr John Fagan and Dr Michael Antoniou received no financial recompense for their work on the project.

About the authors

Claire Robinson, MPhil, is an editor at GMWatch, a public news and information service on issues relating to genetic modification.

John Fagan, PhD was an early voice in the scientific debate on genetically engineered food. Today, as a director of Earth Open Source, Dr Fagan conducts research on biosafety and sustainable agriculture and works to advance environmental sustainability and social responsibility in the food system. Dr Fagan pioneered genetic testing methods for GMOs, and founded, built, and later sold a leading company in this field. Earlier, he researched molecular mechanisms of carcinogenesis at the US National Institutes of Health and in academia. He earned a PhD in biochemistry, molecular biology, and cell biology from Cornell University.

Michael Antoniou, PhD is a Reader in Molecular Genetics and Head of the Gene Expression and Therapy Group, King's College London School of Medicine, UK. He has over 30 years' experience of using genetic engineering technology in the investigation of gene organization and control, with over 50 peer-reviewed publications of original work, and holds inventor status on a number of gene expression biotechnology patents. His discoveries in gene control mechanisms are being used for the production of research, diagnostic and therapeutic products, and safe and efficacious human somatic gene therapy for inherited and acquired genetic disorders.

Contents

Introduction

We began work on the first edition of *GMO Myths and Truths* in 2010, prompted by frequent claims that the case against genetically modifying our food supply had no science behind it. As we had followed the scientific debate and evidence on genetically modified (GM) crops and foods since the early 1990s, we knew that this was untrue.

Another driving factor was the inflated claims that were being made for GM crops. The public was being told that they would make agriculture more sustainable, provide higher yields to feed the world's growing population, reduce pesticide use, help meet the challenges of climate change, provide more nutritious foods, and make farming easier and more profitable.

These claims were at best questionable and at worst patently false. GM has not provided a single crop that has sustainably delivered these benefits. On the contrary, a considerable and growing body of scientific evidence pointed not only to potential hazards but also to actual harm from GMOs (genetically modified organisms) to animal and human health and the environment. But this evidence was not reaching the public, campaigners, policy-makers, or even scientists.

The first edition of *GMO Myths and Truths* was published as a 120-page free download on the Earth Open Source website in 2012. Unexpectedly for such a dry publication, it hit a nerve, coinciding with a push for GMO labelling in the USA. It was also taken up and used by campaigners, policy-makers, and scientists in China, India, South America, Europe, Russia, and Scandinavia.

The second edition was published in 2014 and was three times as long, both to take account of new studies and to answer criticisms from GMO proponents. At the time of writing, *GMO Myths and Truths* has been downloaded half a million times and accessed online by over three times as many visitors.

GMO Myths and Truths: A citizen's guide to the evidence on the safety and efficacy of genetically modified crops and foods is a condensed and updated version for those with limited time and patience. We hope people find it useful.

– Claire Robinson, Michael Antoniou, and John Fagan

1 Myth: Genetic engineering is just an extension of natural breeding and no more risky

Truth: Genetic engineering is radically different from natural breeding and poses special risks

Myth at a glance

Proponents of genetically modified (GM) crops claim genetic engineering is just an extension of natural plant breeding. They say GM is more precise and allows genes coding for the desired trait to be inserted into the host plant with minimal unexpected effects.

But the GM transformation process would never happen in nature. It consists of the random insertion of an artificially constructed foreign gene unit and consequent alteration of the host genome, and plant cell tissue culture.

These processes are not precise but highly mutagenic. They lead to unpredictable changes in the DNA, proteins, and biochemical composition of the resulting GM plant. This can result in the GM plant having unexpected toxic or allergenic effects and altered nutritional value, as well as unpredictable effects on the environment.

Claims that new genome editing techniques, which alter the genetic information of host plant genes, are precise are not supported by evidence. They can have off-target effects in the form of unintended mutations.

Cisgenesis – transferring genes between the same or a closely related species – is claimed to be safer than transgenesis, where genes are moved between different species and/or kingdoms. But the gene cassette used to transfer a cisgene is an artificial construct containing DNA sequences from other species – making the process transgenic. Insertion of the cisgene cassette within the host plant DNA is random and mutagenic. Cisgenesis involves tissue culture, another mutagenic process. Evidence shows that cisgenesis can result in important unanticipated changes to a plant.

Proponents of genetically modified (GM) crops claim genetic engineering is just an extension of natural plant breeding.[1] But GM is technically and conceptually different from natural breeding and poses different risks. This is recognized in national and international laws and agreements on GMOs (genetically modified organisms):

- European law defines a GMO as an organism in which "the genetic material has been altered in a way that does not occur naturally by mating and/or natural recombination" and requires the risks of each GMO to be assessed on a case-by-case basis.[2]
- The Cartagena Protocol on Biosafety,[3] an agreement signed by 168 governments worldwide[4] that seeks to protect biological diversity from the risks of GM technology, and the United Nations food safety body, Codex Alimentarius, agree that GM differs from conventional breeding and that safety assessments should be required before GMOs are used in food or released into the environment.[5,6]

The GMO industry plays both sides in its presentation of GMOs. It tells investors and patent offices that GM is different from natural breeding and involves an "inventive step", thus making GMOs patentable. Yet it tells the public that GM is an extension of natural breeding and that therefore GM foods are as safe as non-GM foods.

Both arguments cannot be correct. And technically speaking, the GM transformation process is radically different from natural breeding.

GM is an artificial laboratory-based technique that is designed to enable the transfer of DNA from any source into any organism. GM even enables the introduction of chemically synthesized DNA into the genome of living organisms. The foreign transgene units transferred by GM are artificially constructed in the laboratory and usually contain components from multiple different organisms. The processes by which the genes are introduced into the organism never happen in nature, but are artificial laboratory processes.

How a GM crop is made

Examination of the steps by which GM crops are created make clear that genetic engineering is not natural and could never occur in nature. Such an examination also shows that contrary to proponents' claims, it is neither precise nor predictable.[7,8]

1. Isolation of genes of interest and cutting and splicing them to generate the GM gene cassette ready for introduction into the plant

Genetic engineering confers one or more traits on an organism by introducing the genes for those traits into the organism's genome. The first step is to identify the gene for a desired trait and isolate it. The isolated gene is then usually propagated as a DNA "clone" in a bacterial host.

Next, pieces of DNA are cut from different isolated genes and spliced together to form a completely novel DNA molecule that never existed before. The resulting molecule will include the DNA sequences that encode the protein for the desired trait, as well as control elements that enable the GM plant to efficiently produce the GM protein. The control elements include sequences that serve as a genetic on-switch (the promoter), enabling the GM gene to be turned on and its information to be used to manufacture the desired GM protein. In addition, essential DNA sequences that mark the end of the GM gene unit (the terminator) are added.

The cutting of the DNA molecules is accomplished using enzymes that recognize specific DNA sequences and cut them out precisely. Then the DNA sequences are spliced together, using another class of enzymes. This process is indeed precise, and GMO proponents point to this step when they want to claim that GM is a precise and controlled process. However, it is the only step in the GM process that can be called precise. Contrary to GMO proponents' claims, the GM process as a whole is highly imprecise.

The result of many cutting and splicing steps is the complete genetically engineered construct, called the GM gene cassette. For example, the GM gene cassette in first-generation GM Roundup Ready soy combines gene sequences from two species of soil bacteria, a flowering plant, and a plant virus. This illustrates the extreme combinations of genetic material that are common in GMOs, but would never occur in nature.

2. GM gene cassette insertion into cultured plant cells

To introduce the GM gene cassette into the genome of the recipient plant, millions of cells from that species are subjected to the GM gene insertion (transformation) process. This is done by growing plant cells or pieces of plant material in dishes under controlled conditions in the laboratory, a system known as "tissue culture". Then the GM gene cassette is inserted into those cells by methods described below. The inserted DNA re-programmes the cells to produce GM proteins, conferring new properties on the cells.

Two methods are commonly used to introduce GM gene cassettes into plant cells. The first, known as biolistics, uses a "gene gun" to shoot microscopic gold or tungsten nanoparticles coated with GM DNA into the plant cells. Millions of cells are bombarded randomly. In a low percentage of cases, the nanoparticles enter the nucleus of plant cells. Once a nanoparticle is inside a cell's, nucleus, the GM DNA can dissolve off the nanoparticle. In an even smaller number of cases, the GM DNA in the nucleus becomes randomly incorporated into the DNA of the plant cell – the aim of the process. This is a random process that genetic engineers do not fully understand and have no ability to control.

The second method of gene insertion is by infection of the cultured plant cells with the soil bacterium *Agrobacterium tumefaciens (A. tumefaciens)*. In its natural form, *A. tumefaciens* infects plants at wound sites, causing grown gall disease, a type of tumour. The infection process involves the actual insertion of DNA from *A. tumefaciens* into the DNA of the infected plant. The genetic engineer uses the natural ability of *A. tumefaciens* to insert DNA into the genome of infected plants in order to insert the GM gene cassette into the DNA of plant cells in culture. This is done by including in the GM gene cassette a large piece of *A. tumefaciens* DNA called the Ti plasmid. This Ti plasmid GM gene cassette is then introduced back into *A. tumefaciens*. When plant cells in culture are infected with *A. tumefaciens* containing the GM gene cassette/Ti plasmid DNA complex, a small fraction of the plant cells become infected and incorporate the GM gene cassette into their own DNA. As with the gene gun insertion process, the *A. tumefaciens* process is hit-or-miss and the genetic engineer has no way of controlling where in the plant cell genome the GM gene cassette will be inserted.

3. Selection of the modified plant cells

At this point in the process, the genetic engineer has a tissue culture consisting of millions of plant cells. Some will have picked up the GM gene cassette, whilst the vast majority will not. Of those that have incorporated the GM gene cassette, only a few will actively express the GM genes contained in it. The genetic engineer now needs to eliminate the cells that do not contain or do not express the GM genes.

To accomplish this, the genetic engineer normally includes a special "selectable marker" gene in the GM gene cassette, along with the gene(s) that confer the GM traits of interest (for example, the insecticidal trait in Bt crops). This additional GM gene expresses a function that enables GM plant cells to be identified and selected for. For instance, marker genes are often used that enable the GM cells to survive in the presence of an antibiotic or herbicide. When the marker gene, along with the other genes in the GM gene cassette, are successfully inserted into the genome of a plant cell and are expressed, the marker gene produces a protein that protects the cell from the antibiotic or herbicide, while the unmodified cells are killed. The genetic engineer can then propagate the surviving cells, all of which will carry the GM gene cassette.

4. Hormone treatment

The few plant cells that have successfully incorporated the GM gene cassette and survived the antibiotic/herbicide treatment are then further treated with plant hormones. These stimulate the genetically modified plant cells to proliferate and differentiate into small GM plants that can be transferred to soil and grown to maturity. Each GM plant produced in this way is unique and has originally arisen from a single GM transformed cell. Thus each will have its own unique range of unpredictability.

5. Verification of the GM transformation

Once the GM plants are growing, the genetic engineer discards any that are deformed or do not grow well. The remaining GM plants carry the same GM gene cassette, but it is inserted at different locations in the plant's genome. The GM plants will express the GM genes at different levels and according to different tissue-specific patterns. The GM plants are tested to identify one or more that express the GM genes at the desired levels and locations in the plant. Out of hundreds or thousands of GM plants produced, only a few may fit this requirement and become candidates for commercialization.

At this stage the GM plants have not been assessed for health and environmental safety or nutritional value. This part of the process is described in later chapters.

The GM transformation process is inefficient

The GM transformation process is a complex multistep process in which each step described above must be successful in order to produce the desired functional GM plant. The randomness of many steps of the process makes it inefficient. In addition, plant cells possess mechanisms for defending themselves against foreign DNA invasion that allows them to degrade or shut down its function and thus block transformation. This makes the overall GM process even more inefficient. Therefore obtaining GM plants that are good candidates for commercialization is a long, labour-intensive, and expensive process.[9,10]

The GM process is highly mutagenic

The process of creating a GMO is highly mutagenic, damaging the sequence of the DNA and altering the genetic information content. Mutations occur in three ways during the GM process, as explained below.[11,12]

1. Insertional mutagenesis caused by insertion of the GM gene cassette

Genetic modification always involves the insertion of a foreign GM gene cassette into the genome (DNA) of the recipient organism. The site of insertion of the foreign GM gene cassette is random. The insertion of the GM gene cassette interrupts the normal sequence of the letters of the genetic code within the DNA of the plant. This is called insertional mutagenesis. If this occurs in the middle of a gene or a regulatory sequence, it will disrupt that function, disturbing the normal expression pattern of the host plant's natural genes. Because insertion occurs at a random location, there is no way of predicting or controlling which of the plant's genes will be disrupted or what effects it will have on the plant.

2. Mutagenesis from secondary insertions

The GM gene insertion process is not clean. Fragments of DNA from the GM gene cassette can insert into the genome of the host plant at multiple random locations. Each of these unintended insertions is a mutational event that can disrupt or destroy the function of other genes or regulatory sequences.

It is estimated that there is a 53–66% probability that any insertional event will disrupt a gene.[11] So if the genetic modification process results in one primary insertion and two or three unintended insertions, it is likely that the function of at least two genes that normally function in the plant will be disrupted.

Research has found that the transformation process can also trigger other kinds of mutations – rearrangements and deletions of the plant's DNA, especially at the site of insertion of the GM gene cassette,[11] which are likely to compromise the functioning of genes important to the plant.

3. Mutations caused by tissue culture

Three steps of the genetic modification process take place while the host plant cells are being grown in tissue culture. These steps include:

→ Initial insertion of the GM gene cassette into the host plant cells

- Selection of plant cells using the selectable marker gene
- Treatment of GM plant cells with hormones, stimulating differentiation into GM plantlets.

The process of tissue culture is obligatory to the genetic engineering process and is in itself highly mutagenic, causing hundreds or thousands of potentially damaging mutations throughout the plant cells' DNA.[11,12]

In the case of plants that are vegetatively propagated (not through seeds but through tubers or cuttings), such as potatoes and bananas, all the mutations resulting from the GM transformation process will be present in the final commercialized crop.

In the case of plants propagated from seed, such as soy and maize, the initial GM plant can be back-crossed multiple times to a non-GM variety. This can "breed out" many, but not all, of the mutations incurred during the GM transformation process. However, since hundreds of genes may be mutated during the GM process, there is a significant risk that mutated versions of multiple genes crucial to important plant functions may remain after this process. This could result in GM crops that are compromised in disease- or pest-resistance, that are allergenic or toxic, or that have unintended alterations in nutritional value.

Can the genetic engineer eliminate harmful mutations?

GM proponents say that even if harmful mutations occur, that is not a problem. They say that during the process of developing a GM crop, the GM plants undergo screening and selection and the genetic engineers will eliminate any that have harmful mutations.[13]

But genetic engineers do not conduct a detailed screening that would catch all plants producing potentially harmful substances. Instead they carry out tests that allow them to identify the few plants, among many thousands, that express the desired trait at the desired level. Of those, they pick some that look healthy and strong to propagate.

Such screening cannot detect subtle changes in the plant's biochemistry arising from mutations caused by the GM process. Yet these mutations can cause plants to produce substances that are harmful to consumers or the environment or to lack important nutrients.

Some agronomic and environmental risks will be missed, as well. For instance, during the GM transformation process, a mutation may destroy a gene that makes the plant resistant to a certain pathogen or a specific environmental stress like extreme heat or drought. But that mutation will be revealed only if the plant is intentionally exposed to the pathogen or stress in a systematic way. GM crop developers are not capable of screening for resistance to every potential pathogen or environmental stress. So mutations can sit like silent time bombs within the GM plant, ready to "explode" at any time when there is an outbreak of the relevant pathogen or an exposure to the relevant environmental stress.

An example of this kind of limitation was an early – but widely planted – variety of Roundup Ready soy. It turned out that this variety was more sensitive than non-GM soy varieties to heat stress and more prone to infection.[14]

Pleiotropic effects

There is another way in which genetic modification generates unintended effects. A single change at the level of the DNA can give rise to multiple changes within the organism.[12,15] Such changes are known as pleiotropic effects. They occur because genes, proteins, and pathways do not act as isolated units but interact with one another and are regulated by a highly complex, multi-layered network of genetic, biochemical and cell biological processes.

Because of these diverse interactions, it is impossible to predict the impacts of even a single GM gene on the organism. The complexity of ecological systems into which the GM crop will be introduced makes it even more challenging to predict the impact of any given GMO.

Unintended changes brought about by the GM process can include alterations in the nutritional content of the food, toxic and allergenic effects, poor crop performance, and the emergence and spread of characteristics that harm the environment. All such unexpected effects have occurred with GMOs, as subsequent chapters show.

It is unlikely that potentially harmful changes would be picked up by the inadequate testing carried out to comply with current government GMO authorization processes. These unexpected changes are especially dangerous because the release of GMOs into the environment is irreversible. Even the worst chemical pollution decreases over time. But GMOs are living organisms that propagate and multiply in the environment, perpetuating any food safety or environmental problems they may cause.

Failed, under-performing, and toxic GM crops

Many GM crops have failed or under-performed under real farming conditions.[14,16,17,18,19,20,21] Not all failures will be due to hidden harmful mutations. Some will be due to environmental effects of the GM crop that are a direct effect of the GM trait expressed as intended. For example, the continuous expression of Bt insecticidal toxins in GM Bt crops creates selective pressure that causes pests to become resistant to the Bt toxins.

Other failures may be due to unexpected interactions between the genome of the plant and environmental factors. For example, the amount of Bt toxin expressed in Bt maize MON810 has been found to be affected by factors such as light[22] and fertilizer use.[23] This type of response could account for some cases in which GM Bt crops have fallen victim to pest attack, since lower Bt expression levels can allow insects to feed on the plant without being killed.

GMO crop failures have mostly been revealed by affected farmers or in tests by non-industry researchers after the crop was commercialized. It seems unlikely that industry carried out thorough "stress tests" in different conditions prior to commercialization – if it did, the results are unpublished.

In the cases of GM foods that have been found to be toxic or allergenic, this could be due to residues of the pesticides associated with the GM crop, or to the GM gene product (for example, Bt toxin). But it could equally be due to one or more mutations in the crop that arose during the GM transformation process. These mutations can cause structural alterations in gene products, or changes in gene expression, or both. They can also cause pleiotropic effects – where the inserted gene or a mutation caused by gene insertion triggers a cascade of events, resulting in multiple unintended gene regulatory or protein/enzyme interactions that cause the plant to produce harmful compounds.

The GM process actively selects for mutational effects in the host genome

The GM process is biased towards damaging important active genes in the plant, because the process selects for insertion of GM gene cassettes into regions of the recipient plant genome that permit it to be switched on and be expressed. These "permissive" regions in the plant DNA are regions where one or more host genes are being actively expressed. These genes are active because they provide important functions useful to the plant. When the GM gene cassette is inserted into a region containing such genes, it is more likely that the insertion will damage one of those functionally important genes, thereby compromising plant function. This can result in reduced viability or tolerance to pests and other stresses, or alterations that cause the GM crop to be toxic, allergenic, or less nutritious, or to have adverse impacts on the environment.

Is GM technology becoming more precise?

Technologies have been developed to target GM gene insertion or sequence modifications to a predetermined site within the plant's DNA in an effort to obtain a more predictable outcome and avoid the complications that can arise from random insertional mutagenesis.[24,25,26,27,28,29]

Some of these technologies use nucleases or "genome scissors" which allow the cutting of DNA and the insertion of new DNA in any position in the chromosomes. The most popular genome scissors are TALENs (transcription activator-like effector nucleases), ZFNs (zinc finger nucleases), and CRISPR-Cas9 (Clustered Regularly Interspaced Short Palindromic Repeats).

Proponents claim that these technologies offer "targeted genome editing".[30] However, these GM transformation methods are not failsafe. All have been found to cause mutations and rearrangements at off-target sites.[31,32,33,34]

These techniques are still new and their effects have not yet been thoroughly researched. Until more experience is gained, we have no basis for assuming that these methods are more precise or less prone to unintended effects.

Biotechnologists still know only a fraction of what there is to be known about the genome of any species and about the genetic, biochemical, and cellular functioning of our crop plants. That means that even if they select an insertion site that they think is safe, insertion of a gene at that site could cause a range of unintended effects, such as disturbances in gene expression or in the function of the protein encoded by that gene.

In addition, the new genome editing methods will still bring about novel combinations of gene functions that carry the risk of unpredictable outcomes at the level of plant biochemistry and composition.

Moreover, because tissue culture must still be carried out for these new targeted insertion methods, the substantial mutagenic effects of the tissue culture process remain a major source of unintended damaging side-effects.

Rapid Trait Development System: GM or not?

The biotechnology companies BASF and Cibus have developed oilseed rape (canola) with a technique called RTDS (Rapid Trait Development System).[35] According to Cibus, RTDS is a method of altering a targeted plant host gene by utilizing the cell's own gene repair system to specifically modify the gene sequence, and does not involve inserting foreign genes or gene expression control sequences.

Cibus markets its RTDS crops as non-transgenic,[35] non-GM,[36] and produced "without the insertion of foreign DNA into plants".[35] The company adds that crops developed using this method are "quicker to market with less regulatory expense".[35] Cibus says that the RTDS method is "all natural", has "none of the health and environmental risks associated with transgenic breeding", and "yields predictable outcomes in plants".[37]

However, GM is a process, and the definition of genetic modification does not depend on the origin of the inserted genetic material. Crops created with RTDS are indeed GMOs, since RTDS alters the genome in a manner that would not occur naturally through breeding or genetic recombination. The fact that no foreign DNA is inserted into the recipient plant's genome is immaterial.

In addition, RTDS still involves tissue culture, which introduces abundant mutations. Some of these mutations (or all, in the case of vegetatively propagated plants, such as potatoes and bananas) will be present in the final marketed product. There will inevitably be off-target effects from the RTDS process. The intent of the RTDS process is specific targeting, but this technique is new and the research has not been done to assess the frequency and extent of off-target effects. The old saying, "Absence of evidence of harm is not evidence of the absence of harm," is pertinent here.

Even changing a single base pair (building block of DNA) in a single gene, whether it encodes an enzyme, a structural protein, a peptide hormone, or a regulatory protein, can cause unintended disturbances in the cell and the organism as a whole. To assess the precision and efficacy of the RTDS process and the extent to which unintended alterations take place at other locations in the genome during RTDS, many studies will be needed.

One important class of studies that must be carried out is whole genome sequencing of RTDS-derived GMOs. Structural and functional analysis of the gene expression profile (transcriptomics) and of the proteins present in these GMOs (proteomics), as well as analysis of metabolites present (metabolomics) would also be required. The agronomic performance, impact on the environment, and quality and safety of RTDS food crops must be investigated, including in long-term toxicological feeding studies – just as with any other GM crop. Today, these data are missing and so many of the claims being made are speculative.

Is cisgenesis a safe form of GM?

Cisgenesis (from "same" and "beginning") is a type of genetic engineering involving artificially transferring genes between organisms from the same species or between closely related organisms that could otherwise be conventionally bred. For example, a cisgenic GM potato engineered to resist blight was developed using a gene taken from a wild potato.[38]

Proponents claim that cisgenesis is safer than transgenesis, as purportedly it involves transfer of genetic material only between members of the same species and no foreign genes are introduced.[39,40] Some scientists are calling for deregulation of cisgenic plants on the grounds that they carry no additional risks to those of naturally bred plants.[41,42,43]

However, cisgenesis still carries most of the risks associated with transgenic genetic engineering, for the following reasons:

- No GMO called cisgenic is purely cisgenic. The word "cisgenic" (meaning "same descent") implies that only DNA from the genome of the same or closely related species are being introduced. But no supposedly cisgenic GMO has ever been or is likely to be created using only DNA from its own species. Although it is possible to isolate a gene from maize, for instance, and then put it back into maize, this will not be a purely cisgenic process. In order to put the gene back into maize, it is necessary to link it to other sequences, at least from bacteria, and possibly also from viruses, other organisms (potentially from different species), and even synthetic DNA.[44,45] So "cisgenic" actually means "partly transgenic". For example, the cisgenic plants engineered by scientists who claimed to have made "the first genetically engineered plants that contain only native DNA" were produced using genetic modification mediated by the soil bacterium *A. tumefaciens* – an organism from a different species. The final supposedly cisgenic plant contained GM sequences from *A. tumefaciens*, and was therefore actually transgenic.[46]
- Cisgenic GMOs use the same mutagenic transformation techniques as transgenic GMOs.[12] The process of inserting any fragment of DNA, whether cisgenic or transgenic, into an organism via the GM transformation process using the common *Agrobacterium* or biolistics methods carries the risks associated with insertional mutagenesis. Insertion takes place in an uncontrolled manner and results in at least one insertional mutation event within the DNA of the recipient

organism. The insertional event will interrupt some sequence within the DNA of the organism and may interfere with any natural function that the interrupted DNA carries.

- Cisgenesis, like transgenic genetic engineering, invariably involves the tissue culture process, which has wide-scale mutagenic effects on the plant host DNA.

Cisgenesis can be as unpredictable as transgenesis

Claims that cisgenesis is safe and predictable have been thrown into question by experiments using the model plant *Arabidopsis thaliana*.[48,49,50,51] The results showed that trait introduction via a cisgene can result in plants that differ in unanticipated and dramatic ways from their conventionally bred counterparts. The differences observed have important agronomic and ecological implications.[45] They included higher levels of outcrossing[50] and decreased fitness[48,49] in cisgenic compared with conventionally bred plants.

References

1. GeneWatch UK. ASA rules that Monsanto adverts were misleading: GeneWatch UK complaints upheld [press release]. http://www.genewatch.org/article.shtml?als[cid]=492860&als[itemid]=507856. Published August 10, 1999.
2. European Parliament and Council. Directive 2001/18/EC of the European Parliament and of the Council of 12 March 2001 on the deliberate release into the environment of genetically modified organisms and repealing Council Directive 90/220/EEC. Off J Eur Communities. 2001:1-38.
3. Secretariat of the Convention on Biological Diversity. Cartagena Protocol on Biosafety to the Convention on Biological Diversity. Montreal; 2000. http://bch.cbd.int/protocol/text/.
4. United Nations Convention on Biological Diversity. List of parties: Cartagena Protocol. 2014. http://www.cbd.int/information/parties.shtml#tab=1.
5. Codex Alimentarius. Foods Derived from Modern Biotechnology (2nd Ed.). Rome, Italy: World Health Organization/Food and Agriculture Organization of the United Nations; 2009. ftp://ftp.fao.org/codex/Publications/Booklets/Biotech/Biotech_2009e.pdf.
6. Codex Alimentarius. Guideline for the Conduct of Food Safety Assessment of Foods Derived from Recombinant-DNA Plants: CAC/GL 45-2003; 2003.
7. Bawden T. GM crops are safer than conventional crops, says Environment Secretary Owen Paterson. The Independent. http://www.independent.co.uk/life-style/food-and-drink/news/gm-crops-are-safer-than-conventional-crops-says-environment-secretary-owen-paterson-8665872.html. Published June 20, 2013.
8. Rush T. Europe "must switch to GM for precision." Farmers Guardian. http://www.farmersguardian.com/home/arable/arable-news/europe-%E2%80%98must-switch-to-gm-for-precision%E2%80%99/28657.article. Published November 6, 2009.
9. Phillips McDougall. The Cost and Time Involved in the Discovery, Development and Authorisation of a New Plant Biotechnology Derived Trait: A Consultancy Study for Crop Life International. Pathhead, Midlothian; 2011.
10. Goodman MM. New sources of germplasm: Lines, transgenes, and breeders. In: Martinez JM, ed. Memoria Congresso Nacional de Fitogenetica. Univ Autonimo Agr Antonio Narro, Saltillo, Coah, Mexico; 2002:28-41. http://www.cropsci.ncsu.edu/maize/publications/NewSources.pdf.
11. Latham JR, Wilson AK, Steinbrecher RA. The mutational consequences of plant transformation. J Biomed Biotechnol. 2006;2006:1-7. doi:10.1155/JBB/2006/25376.
12. Wilson AK, Latham JR, Steinbrecher RA. Transformation-induced mutations in transgenic plants: Analysis and biosafety implications. Biotechnol Genet Eng Rev. 2006;23:209-238.
13. Academics Review. The use of tissue culture in plant breeding is not new. 2014. http://bit.ly/

I7fPc9.
14. Coghlan A. Monsanto's Roundup-Ready soy beans cracking up. New Sci. 1999. http://www.biosafety-info.net/article.php?aid=250.
15. Pusztai A, Bardocz S, Ewen SWB. Genetically modified foods: Potential human health effects. In: D'Mello JPF, ed. Food Safety: Contaminants and Toxins. Wallingford, Oxon: CABI Publishing; 2003:347-372. http://www.leopold.iastate.edu/sites/default/files/events/Chapter16.pdf.
16. Stauffer C. UPDATE 1-Brazil farmers say GMO corn no longer resistant to pests. Reuters. http://www.reuters.com/article/2014/07/28/brazil-corn-pests-idUSL2N0Q327P20140728. Published July 28, 2014.
17. MASIPAG (Philippines). 10 Years of Failure, Farmers Deceived by GM Corn.; 2013. https://www.youtube.com/watch?v=hCuWs8K9-kI&list=LL6feir9VTqIG5CV3-V29IcQ&feature=mh_lolz.
18. Patel R. Making up Makhatini. In: Stuffed and Starved. London, UK: Portobello Books; 2007:153-158.
19. Hammadi S. Bangladeshi farmers caught in row over $600,000 GM aubergine trial. The Guardian. http://www.theguardian.com/environment/2014/jun/05/gm-crop-bangladesh-bt-brinjal. Published June 5, 2014.
20. Gillam C. Scientists warn EPA on Monsanto corn rootworm. Reuters. http://www.reuters.com/article/2012/03/09/us-monsanto-corn-idUSBRE82815Z20120309. Published March 9, 2012.
21. Gurian-Sherman D. Failure to Yield: Evaluating the Performance of Genetically Engineered Crops. Cambridge, MA: Union of Concerned Scientists; 2009. http://www.ucsusa.org/assets/documents/food_and_agriculture/failure-to-yield.pdf.
22. Abel CA, Adamczyk JJ. Relative concentration of Cry1A in maize leaves and cotton bolls with diverse chlorophyll content and corresponding larval development of fall armyworm (Lepidoptera: Noctuidae) and southwestern corn borer (Lepidoptera: Crambidae) on maize whorl leaf profiles. J Econ Entomol. 2004;97(5):1737-1744.
23. Bruns HA, Abel CA. Effects of nitrogen fertility on Bt endotoxin levels in maize. J Entomol Sci. 2007;42(1):35-43.
24. Kumar S, Fladung M. Controlling transgene integration in plants. Trends Plant Sci. 2001;6:155-159.
25. Ow DW. Recombinase-directed plant transformation for the post-genomic era. Plant Mol Biol. 2002;48:183-200.
26. Li Z, Moon BP, Xing A, et al. Stacking multiple transgenes at a selected genomic site via repeated recombinase-mediated DNA cassette exchanges. Plant Physiol. 2010;154:622-631. doi:10.1104/pp.110.160093.
27. Shukla VK, Doyon Y, Miller JC, et al. Precise genome modification in the crop species Zea mays using zinc-finger nucleases. Nature. 2009;459(7245):437-441. doi:10.1038/nature07992.
28. Townsend JA, Wright DA, Winfrey RJ, et al. High-frequency modification of plant genes using engineered zinc-finger nucleases. Nature. 2009;459(7245):442-445. doi:10.1038/nature07845.
29. Shen H. CRISPR technology leaps from lab to industry. Nature. 2013. doi:10.1038/nature.2013.14299.
30. Wood AJ, Lo T-W, Zeitler B, et al. Targeted genome editing across species using ZFNs and TALENs. Science. 2011;333(6040):307. doi:10.1126/science.1207773.
31. Mussolino C, Morbitzer R, Lutge F, Dannemann N, Lahaye T, Cathomen T. A novel TALE nuclease scaffold enables high genome editing activity in combination with low toxicity. Nucleic Acids Res. 2011;39(21):9283-9293. doi:10.1093/nar/gkr597.
32. Pattanayak V, Ramirez CL, Joung JK, Liu DR. Revealing off-target cleavage specificities of zinc-finger nucleases by in vitro selection. Nat Methods. 2011;8(9):765-770. doi:10.1038/nmeth.1670.
33. Gabriel R, Lombardo A, Arens A, et al. An unbiased genome-wide analysis of zinc-finger nuclease specificity. Nat Biotechnol. 2011;29(9):816-823. doi:10.1038/nbt.1948.
34. Fu Y, Foden JA, Khayter C, et al. High-frequency off-target mutagenesis induced by CRISPR-Cas nucleases in human cells. Nat Biotechnol. 2013;31(9):822-826. doi:10.1038/nbt.2623.
35. Cibus. BASF and Cibus achieve development milestone in CLEARFIELD® production system [press release]. Undated. http://www.cibus.com/press/press012709.php.
36. Cibus. The Rapid Trait Development System. 2014. http://cibus.com/rtds.php.
37. Cibus. The evolution of plant breeding: RTDS™ versus other technologies. Undated. http://www.cibus.com/pdfs/RtdsSketch4_LoRes.pdf.
38. Jones JDG, Witek K, Verweij W, et al. Elevating crop disease resistance with cloned genes. Philos Trans R Soc B Biol Sci. 2014;369(1639):20130087. doi:10.1098/rstb.2013.0087.
39. Rommens CM. Intragenic crop improvement: Combining the benefits of traditional breeding

and genetic engineering. J Agric Food Chem. 2007;55:4281-4288. doi:10.1021/jf0706631.
40. Rommens CM, Haring MA, Swords K, Davies HV, Belknap WR. The intragenic approach as a new extension to traditional plant breeding. Trends Plant Sci. 2007;12:397-403. doi:10.1016/j.tplants.2007.08.001.
41. Schouten HJ, Krens FA, Jacobsen E. Cisgenic plants are similar to traditionally bred plants. EMBO Rep. 2006;7(8):750-753. doi:10.1038/sj.embor.7400769.
42. Schouten HJ, Krens FA, Jacobsen E. Do cisgenic plants warrant less stringent oversight? Nat Biotechnol. 2006;24(7):753-753. doi:10.1038/nbt0706-753.
43. Viswanath V, Strauss SH. Modifying plant growth the cisgenic way. ISB News. 2010.
44. Rommens CM. All-native DNA transformation: a new approach to plant genetic engineering. Trends Plant Sci. 2004;9(9):457-464. doi:10.1016/j.tplants.2004.07.001.
45. Wilson A, Latham J. Cisgenic plants: Just Schouten from the hip? Indep Sci News. 2007. http://www.independentsciencenews.org/health/cisgenic-plants/.
46. Rommens CM, Humara JM, Ye J, et al. Crop improvement through modification of the plant's own genome. Plant Physiol. 2004;135(1):421-431. doi:10.1104/pp.104.040949.
47. Schubert D, Williams D. "Cisgenic" as a product designation. Nat Biotechnol. 2006;24(11):1327-1329. doi:10.1038/nbt1106-1327.
48. Bergelson J, Purrington CB, Palm CJ, Lopez-Gutierrez JC. Costs of resistance: A test using transgenic Arabidopsis thaliana. Proc Biol Sci. 1996;263:1659-1663. doi:10.1098/rspb.1996.0242.
49. Purrington CB, Bergelson J. Fitness consequences of genetically engineered herbicide and antibiotic resistance in Arabidopsis thaliana. Genetics. 1997;145(3):807-814.
50. Bergelson J, Purrington CB, Wichmann G. Promiscuity in transgenic plants. Nature. 1998;395:25. doi:10.1038/25626.
51. Bergelson J, Purrington C. Factors affecting the spread of resistant Arabidopsis thaliana populations. In: Letourneau D, Elpern Burrows B, eds. Genetically Engineered Organisms. CRC Press; 2001:17-31. http://www.crcnetbase.com/doi/abs/10.1201/9781420042030.ch2.

2 Myth: GM foods are strictly tested and regulated for safety

Truth: Regulation relies on GM food safety tests conducted by developer companies and regulatory processes are weak

Myth at a glance

Claims that GM foods are extensively tested and strictly regulated are false. At best, they are tested for safety prior to commercialization by the companies that stand to profit from selling them. The tests are weak and inadequate to show safety, especially regarding the long-term consumption of these products.

GM foods were first allowed into the human food supply in the US, based on the claim that they are Generally Recognized as Safe (GRAS) – but no GM food has ever fulfilled the strict GRAS criteria.

In many countries, GM foods are approved by regulators as "substantially equivalent" to non-GM crops. But when this assumption is tested scientifically, GM crops are often found to have unexpected and unintended differences. These differences may make the GM food more toxic or allergenic than the non-GM parent variety.

Industry and some government sources claim that GM foods are strictly regulated.[1,2] But GM food regulatory systems worldwide vary from extremely weak (in the US) to inadequate (in Europe). None are adequate to protect consumers' health. All regulators rely on safety testing done by the GMO crop developer company that wishes to commercialize the GMO in question.

As criticism has mounted of the deficiencies in GM food regulatory systems, the message from pro-GM lobbyists has shifted, from "GM foods are strictly regulated" to "GM foods are no more risky than non-GM foods, so why regulate them at all?" They argue that each time a plant breeder develops a new variety of apple or beetroot through conventional breeding, we do not demand that it be tested toxicologically, and there is no reason to think that GM foods will be any more toxic.

But this argument is spurious. Genetic engineering, including so-called "cisgenesis", uses genes from widely differing and unrelated organisms – unlike conventional breeding, where the organisms to be bred must be closely related. In addition, conventional breeding uses natural reproductive mechanisms that have evolved to be precise and that have built in "proof-reading" mechanisms that catch many errors/mutations that do occur. And humans have co-evolved with their food crops over millennia and have learned by long – and doubtless sometimes bitter – experience which plants are toxic and which are safe to eat. Any unexpected cases of harm would have been localized in effect and the toxic plants selected out of future breeding programmes.

With GM foods, we do not have the luxury of long periods of experimentation by our ancestors. And unlike our ancestors, we show no sign of learning from the mistakes of genetic engineering, since signs of toxicity in animal feeding experiments with GM foods are routinely dismissed (see Chapter 4).

Regulatory process is based on industry studies

> *"One thing that surprised us is that US regulators rely almost exclusively on information provided by the biotech crop developer, and those data are not published in journals or subjected to peer review... The picture that emerges from our study of US regulation of GM foods is a rubber-stamp 'approval process' designed to increase public confidence in, but not ensure the safety of, genetically engineered foods."*
>
> – David Schubert, professor and head, Cellular Neurobiology Laboratory, Salk Institute, USA[3,4]

Worldwide, government regulatory agencies decide on the safety of a GMO based on studies commissioned and controlled by the same companies that stand to profit from the crop's approval. The problem with this system is that industry studies are biased. Published reviews of studies assessing the risks of controversial products and technologies show that industry-sponsored studies, or studies where authors are affiliated with industry, are much more likely to reach a favourable conclusion about the safety of the product than studies carried out by scientists independent of industry. This applies to studies on smoking,[5,6] pharmaceuticals,[7,8] medical products,[9] and mobile phone technology.[10]

GMOs are no exception. A review of scientific studies on the health risks of GM crops and foods showed that either financial or professional conflict of interest (author affiliation to industry) was strongly associated with study outcomes that cast GM products in a favourable light.[11]

How GMOs first entered world markets

GM foods were first commercialized in the US in the early 1990s. The US Food and Drug Administration (FDA) allowed GM foods onto world markets in spite of its own scientists' warnings that genetic engineering is different from conventional breeding and poses special risks, including the production of new toxins or allergens that are difficult to detect.[12,13,14,15,16,17]

For example, FDA microbiologist Dr Louis Pribyl stated: "There is a profound difference between the types of unexpected effects from traditional breeding and genetic engineering". He added that several aspects of genetic engineering "may be more hazardous".[17]

FDA official Linda Kahl protested that the agency was "trying to fit a square peg into a round hole" by "trying to force an ultimate conclusion that there is no difference between foods modified by genetic engineering and foods modified by traditional breeding practices." Kahl stated: "The processes of genetic engineering and traditional breeding are different, and... lead to different risks."[12]

Several FDA scientists called for more rigorous scientific data to be presented by the companies before GMOs were released onto the market, and specifically for safety and toxicological testing.[13,14,17] But FDA administrators, who admitted that the agency was following a government agenda to "foster" the growth of the biotech industry,[18] disregarded their scientists' concerns and permitted GMOs to enter the market without meaningful testing or labelling.

The creation of this policy was overseen by the FDA's deputy commissioner of policy, Michael Taylor, who was appointed to the post in 1991. Prior to joining the FDA, Taylor had been Monsanto's attorney. In 1998 he became Monsanto's vice president for public policy.[19,20] By 2010 he was back at the FDA as its deputy commissioner for foods.[21]

Taylor's career is often cited as an example of a type of conflict of interest known as the "revolving door". The term describes the movement of personnel between regulatory agencies and the industries they regulate.

The US regulatory process for GMOs

The US FDA does not have a mandatory GM food safety assessment process and has never approved as safe any GM food that is currently on the market.

It does not carry out or commission safety tests on GM foods. Instead, the FDA conducts a voluntary pre-market review of a *summary* of the research the manufacturer has conducted.

Although all GM foods commercialized to date have gone through this lenient and superficial process, there is no legal requirement for them to do so. The outcome of the FDA's voluntary assessment is not a conclusion, underwritten by the FDA, that the GMO is safe. Instead it consists of the FDA sending the company a letter stating that:

- The company has provided the FDA with a summary of research that it has conducted assessing the GM crop's safety
- Based on the results of the research done by the company, the company has concluded that the GMO is safe
- The FDA has no further questions
- The company is responsible for placing only safe foods in the market
- If a product is found to be unsafe, the company may be held liable.[22]

This process does not guarantee – or even attempt to scientifically investigate – the safety of GM foods. Though it may protect the image of GM foods, it does not protect the public.

Biotechnology Consultation Agency Response Letter BNF No. 000001

Return to inventory: Completed Consultations on Foods from Genetically Engineered Plant Varieties

See also Biotechnology: Genetically Engineered Plants for Food and Feed and about Submissions on Bioengineered New Plant Varieties

See FDA's memo on BNF No. 000001 for further details

January 27, 1995

Ms. Diane Re
Regulatory Affairs
The Agricultural Group of Monsanto
700 Chesterfield Parkway North
Chesterfield, MO 63198

Dear Ms. Re:

This is in regard to your genetically modified glyphosate-tolerant soybean about which you initiated consultations with the agency in June 1993. The new soybean variety has been rendered tolerant to glyphosate by expression of a 5-enolpyruvylshikimate-3-phosphate synthase from the bacterium *Agrobacterium sp.* strain CP4.

As part of bringing your consultation with FDA regarding this product to closure, you submitted a summary of your safety and nutritional assessment of the new soybean variety on September 2, 1994. On September 19, 1994, you also made a detailed oral presentation of the data that support your submission. It is our understanding that these communications were intended by Monsanto to inform FDA of the steps taken to ensure that this product complies with the Federal Food, Drug, and Cosmetic Act. Further, it is our understanding that, based on the safety and nutritional assessment you have conducted, you have concluded that the new soybean variety is not materially different in composition, safety, or any other relevant parameter from soybean varieties currently on the market and that it does not raise issues that would require premarket review or approval. All materials relevant to this consultation have been placed in a file that has been designated BNF 0001 and that will be maintained in the Office of Premarket Approval.

Based on the description of the data and information presented during the consultations, the new soybean variety does not appear to be significantly altered within the meaning of 21 CFR 170.30(f)(2). We have no additional questions concerning the product at this time. However, as you are aware, it is Monsanto's continued responsibility to ensure that foods the firm markets are safe, wholesome, and in compliance with all applicable legal and regulatory requirements.

Sincerely yours,

Alan M. Rulis, Ph.D.
Acting Director
Office of Premarket Approval
Center for Food Safety
and Applied Nutrition

Letter from the US FDA's Center for Food Safety and Applied Nutrition to Monsanto regarding its GM glyphosate-tolerant soybean. The letter confirms that the FDA is not liable if any safety concerns are identified with the soybean.

The US government is not impartial regarding GM crops and foods

The US government is not an impartial authority but a promoter of GM crops. Through its embassies and agencies, the US government pressures other governments to accept GM crops. Diplomatic cables disclosed by Wikileaks revealed that:

- The US embassy in Paris recommended that the US government launch a retaliation strategy against the EU that "causes some pain" as punishment for Europe's reluctance to adopt GM crops.[23]
- The US embassy in Spain suggested that the US government and Spain draw up a joint strategy to help boost the development of GM crops in Europe.[24]
- The US State Department is trying to steer African countries towards acceptance of GM crops.[25,26]

Who is responsible for ensuring the safety of GM food?

"Monsanto should not have to vouchsafe the safety of biotech food. Our interest is in selling as much of it as possible. Assuring its safety is the FDA's job."

– Philip Angell, Monsanto's director of corporate communications (the FDA is the US government's Food and Drug Administration, responsible for food safety)[27]

"Ultimately, it is the food producer who is responsible for assuring safety."

– US Food and Drug Administration (FDA)[28]

"It is not foreseen that EFSA carry out such [safety] studies as the onus is on the [GM industry] applicant to demonstrate the safety of the GM product in question."

– European Food Safety Authority (EFSA)[29]

FDA presumes GMOs are "generally recognized as safe"

The US FDA holds that GM foods can be marketed without prior testing or oversight because they are "generally recognized as safe" or GRAS.[28] According to US statutory law and FDA regulations, a food that does not have a history of safe use prior to 1958 must meet the following criteria in order to qualify as GRAS:

- There is an overwhelming expert consensus that it is safe; and
- This consensus is based on scientific evidence generated through "scientific procedures", which "shall ordinarily be based upon published studies".[30]

GM foods have never met either requirement and should not be classified as GRAS. At the time the FDA made its presumption that GM foods are GRAS, there was not even expert consensus about their safety within the FDA (as attested by the statements of the agency's scientists detailed above). The FDA's biotechnology coordinator admitted there was no such consensus outside the agency either.[31]

No scientific consensus on GMO safety has emerged since then. In 2001 an expert panel of the Royal Society of Canada declared that it is "scientifically unjustifiable" to presume that GM foods are safe.[32] In 2014 over 300 scientists signed a statement asserting that there was no scientific consensus on GMO safety and that claims of a consensus in the media were an "artificial construct that has been falsely perpetuated".[33]

The sham of substantial equivalence

"The concept of substantial equivalence has never been properly defined; the degree of difference between a natural food and its GM alternative before its 'substance' ceases to be acceptably 'equivalent' is not defined anywhere, nor has an exact definition been agreed by legislators. It is exactly this vagueness that makes the concept useful to industry but unacceptable to the consumer... Substantial equivalence is a pseudo-scientific concept because it is a commercial and political judgment masquerading as if it were scientific. It is, moreover, inherently anti-scientific because it was created primarily to provide an excuse for not requiring biochemical or toxicological tests."

– Erik Millstone, professor of science and technology policy, University of Sussex, UK, and colleagues[34]

"Substantial equivalence is a scam. People say that a potato has vaguely the same amount of protein and starch and stuff as all other potatoes, and therefore that it is substantially equivalent, but that is not a test of anything biological."

– Professor C. Vyvyan Howard, medically qualified toxicopathologist, then at the University of Liverpool, UK[35]

> *"In one interpretation, to say that the new [GM] food is 'substantially equivalent' is to say that 'on its face' it is equivalent (i.e. it looks like a duck and it quacks like a duck, therefore we assume that it must be a duck – or at least we will treat it as a duck). Because 'on its face' the new food appears equivalent, there is no need to subject it to a full risk assessment to confirm our assumption... 'Substantial equivalence' does not function as a scientific basis for the application of a safety standard, but rather as a decision procedure for facilitating the passage of new products, GM and non-GM, through the regulatory process."*
>
> –Royal Society of Canada[32]

Worldwide, regulators approve GM foods as safe based on the concept of "substantial equivalence". Substantial equivalence assumes that if a GMO contains similar amounts of a few basic components, such as protein, fat, and carbohydrate, as its non-GM counterpart, then the GMO is substantially equivalent to the non-GMO and no rigorous safety testing is required.

The concept of substantial equivalence as applied to GMOs was first put forward by the industry and the Organization for Economic Cooperation and Development (OECD), a body dedicated not to protecting public health but to facilitating international trade.[36,37]

Until recently there has been no legal or scientific definition of substantial equivalence. It had not been established how different a GM crop is allowed to be in its constituents from the non-GM parent line or other varieties of the crop, before it is deemed non-substantially equivalent.[38] In 2013, after years of criticism, the EU defined limits on the extent to which a GMO can differ from the non-GM comparator and still qualify as equivalent.[39] However, the limits are set so wide that as long as the measured components in the GM crop fall within the "range of natural variation" of crops with a history of safe use and are not expected to be "biologically relevant",[39] no regulatory action is likely to be triggered.

Claims of substantial equivalence for GM foods have been criticized and revealed as scientifically inaccurate by independent researchers.[40,41,42,43,32]

Technically, the issue is that it is not the total gross amount of protein, carbohydrate or fat that is important, but rather the spectrum of different proteins, carbohydrates or fats that constitute these substances. For substantial equivalence to be scientifically meaningful, a GM crop would not only need to have roughly the same *total* amount of protein but also to have the same levels of each *individual* protein that is present in the non-GM parent crop.

Significant differences in individual protein, carbohydrate or fat components in the GM crop compared to its non-GM parent indicate disrupted gene function and biochemistry stemming from the GM process and thus make the GM food *not* substantially equivalent.

The technology to conduct a detailed "molecular profiling" of foods is readily available and easy to use. Regrettably, however, regulators do not require, and indeed actively resist, the use of this state-of-the-art molecular profiling analytical technology as part of the safety evaluation of GM foods. Perhaps this is due to the fact that, as illustrated below, GM crops are revealed to be *not* substantially equivalent to non-GM crops when molecular profiling is used to compare them. They are invariably found to have significant compositional differences – differences that should require regulators to request major additional toxicity evaluations.

BSE-infected cow: Substantially equivalent to a healthy cow?

A useful analogy to help us understand what is meant by substantial equivalence is that of a BSE-infected cow and a healthy cow. They are substantially equivalent to one another, in that their chemical composition is the same. The only difference is in the shape of one protein (prion) that constitutes a minute proportion of the total mass of the cow. This difference would not be picked up by current substantial equivalence assessments. Yet few would claim that eating a BSE-infected cow is as safe as eating a healthy cow.

GM crops are not substantially equivalent

When claims of substantial equivalence are tested, they are often found to be untrue. Using molecular analytical methods, GM crops have been shown to have a different composition to their non-GM counterparts. This is true even when the two crops are grown under the same conditions, at the same time and in the same location. This means that the changes are not due to different environmental factors but to the genetic modification. The important factor is that these changes should not have occurred, since GM crops are approved on the basis that they are substantially equivalent to non-GM crops except for the deliberately added trait, for example, the Bt toxin insecticidal trait.

Examples of unintended compositional changes in GM crops include:

- GM soy had 12–14% lower amounts of isoflavones, compounds that play a role in sex hormone metabolism, than non-GM soy.[44]

- GM soy had 27% higher levels of a major allergen, trypsin-inhibitor, than the non-GM parent variety, despite the Monsanto authors' claim that the GM soybean was "equivalent" to the non-GM soybean. In order to reach the conclusion of "equivalence", the Monsanto authors compared plants grown at different locations and different times, increasing the range of variability with irrelevant data. Good scientific practice in a test of substantial equivalence requires the GM plant to be compared with the non-GM isogenic (with the same genetic background) variety, grown at the same time in the same conditions.[45]
- Canola (oilseed rape) engineered to contain vitamin A in its oil had much reduced vitamin E and an altered oil-fat composition, compared with non-GM canola.[46]
- Experimental GM rice varieties had unintended major nutritional disturbances compared with non-GM counterparts, although they were grown side-by-side in the same conditions. The structure and texture of the GM rice grain was affected and its nutritional content and value were dramatically altered. The authors said that their findings provided "alarming information with regard to the nutritional value of transgenic rice" and showed that the GM rice was not substantially equivalent to non-GM.[47]
- GM Bt insecticidal maize MON810 had a markedly altered protein profile compared with that of the non-GM counterpart. Unexpected changes included the appearance of a new form of the protein zein, a known allergen, which was not present in the non-GM maize. Other proteins were present in both their natural forms and in truncated and lower molecular mass forms.[43] These findings suggest major disruptions in gene structure and function in this GM crop, which could potentially cause problems of unexpected allergenicity or toxicity.
- GM Bt maize variety MON810 Ajeeb YG showed significant differences from its isogenic non-GM counterpart, with some values being outside the range recorded in the scientific literature. Some fatty acids and amino acids present in the non-GM maize were absent in the Bt maize. The researchers concluded that the genetic modification process had caused alterations in the maize that could result in toxicity to humans and animals.[48]

Altered nutritional value is of concern for two reasons:

1. It could directly affect the health of the human or animal consumer by providing an excess or shortage of certain nutrients.
2. It is an indicator that the genetic engineering process could have

altered biochemical processes in the plant. This could signify that other unexpected changes have also occurred that might impact human or animal health, such as altered toxicity or allergenicity. Indeed, the Bt maize MON810 Ajeeb YG and its non-GM counterpart, which were found to be compositionally different,[48] were tested in a rat feeding study and the GM variety was found to cause toxicity to multiple organs.[49,50]

Europe's comparative safety assessment: Substantial equivalence by another name

Europe has adopted the concept of substantial equivalence in its GM food assessments – but under another name. The European Food Safety Authority (EFSA) does not use the discredited term "substantial equivalence" but has allowed industry to replace it with another term with the same meaning: "comparative assessment" or "comparative safety assessment".

The story of how the comparative safety assessment made its way into Europe's GMO regulatory system is, like the formation of the US FDA's biotechnology policy, a tale of revolving doors and conflicts of interest with industry.

The change of name from "substantial equivalence" to "comparative safety assessment" was suggested in a 2003 paper on risk assessment of GM plants.[51] The paper was co-authored by Harry Kuiper, then chair of EFSA's GMO Panel, with Esther Kok. In 2010 Kok joined EFSA as an expert on GMO risk assessment.[52] In their 2003 paper, Kuiper and Kok admitted that the concept of substantial equivalence remained unchanged and that the name change was in part meant to deflect the "controversy" around the term.[51]

At the same time that Kuiper and Kok published their 2003 paper, they were part of a task force of the GMO industry-funded International Life Sciences Institute (ILSI), that was working on re-designing GMO risk assessment.[37] In 2004 Kuiper and Kok co-authored an ILSI paper which defines comparative safety assessment. The other authors include representatives of GM crop companies that sponsor ILSI, including Monsanto, Bayer, Dow, and Syngenta.[53]

The right and wrong way to do a comparative assessment

The proper scientific method of carrying out a comparative assessment is to grow the GM crop and its non-GM isogenic comparator side-by-side under the same conditions. This ensures that any differences found in the

GM crop are known to arise from the genetic modification and not from environmental factors such as different growing conditions. It also fulfills the intent of European GMO legislation, which is to enable differences "arising from the genetic modification" to be identified and assessed.[54]

If differences are found between the GM crop and the correct comparator, this is a sign that the genetic engineering process has caused disruption in the structure and/or function of the native genes of the host plant. Further investigations should then be carried out to look for other unintended changes.

In contrast, comparisons with unrelated or distantly related varieties grown at different times and in different locations introduce and increase external variables and serve to mask rather than highlight the effects of the genetic engineering process.

This, however, is the method favoured by the GMO industry, both in the compositional analyses it performs on its products[55,56] and in the animal feeding trials it carries out on its GMOs for regulatory authorizations. In its animal feeding trials, it compares the GMO diet not only with the non-GM isogenic comparator diet, but also with a range of "reference" diets containing varieties grown in different locations.[57,58] The effect is to hide the effects of the genetic modification on the plant amid the "noise" created by the external variables.

This is a misuse of the concept of comparative assessment or substantial equivalence. Assessing substantial equivalence or carrying out a comparative assessment is not a safety assessment in itself. It was originally designed to determine whether deeper assessment of the safety of the GMO is needed. A determination that the GMO is not substantially equivalent does not mean that it is hazardous, but indicates that the GM process has altered the functioning of the organism and that deeper assessment is needed to determine whether those alterations have resulted in changes that make the GMO harmful to health or the environment.

The parameters measured are not intended to be indicators of hazard or safety but are indicators of change in function due to genetic modification. They show if it is necessary to look more carefully at whether the changes have resulted in a safe or harmful crop.

From this perspective, it does not matter whether the measured value of a parameter is "within the range of normal variation" or not. What matters is that the value is or is not significantly different from the value for the non-GMO isogenic control. If it is different, this is an indicator that the gene expression, metabolic functioning, and/or cellular regulation of the GMO is significantly different from that of the non-GMO isogenic control and that it is necessary to investigate further to understand whether the differences have resulted in a crop that is safe or dangerous.

GMOs would not pass an objective comparative safety assessment

If the comparative safety assessment were applied objectively and systematically with proper controls, most GMOs would not pass even this weak test of safety. This is because as explained above ("The sham of substantial equivalence"), many studies on GM crops show that they are not substantially equivalent to the non-GM counterparts from which they are derived. There are often significant differences in the levels of certain nutrients and types of proteins, which could impact allergenicity, toxicity, or nutritional value.

The GMO industry and its supporters have sidestepped this problem by widening the range of comparison. Adopting a method used by Monsanto in analyses of its GM soy,[55,56] they no longer restrict the comparator to the GM plant and the genetically similar (isogenic) non-GM line, grown side-by-side under the same conditions and at the same time. Instead they use as comparators a range of non-isogenic varieties grown at different times and in different locations. Some of this "historical" data even dates back to before World War II.[59] It is likely to have been generated by different researchers using methods that differ in sensitivity, accuracy and reliability, relative to the methods used by the authors of the comparative assessment. Anyone familiar with basic principles of experimental design will recognize that comparisons to such data are not meaningful.

This strategy subverts the whole concept of substantial equivalence and comparative assessment – from being a preliminary method of assessing whether the internal functioning of the GMO is different from the original crop, to a safety assessment in and of itself. Showing that the GMO is within the range of normal variation is wrongly used as evidence that the GMO crop is safe.

Despite the loose approach taken in these comparative assessments, they often still reveal statistically significant differences in composition between the GMO and the diverse comparator dataset used by the company applying for approval of the GMO. In other words, the properties of the GMO are outside the range of the non-GMO comparator data, including the historical data. But even in these extreme cases, the differences are dismissed as not being "biologically relevant".[59]

The ILSI database

The industry-funded International Life Sciences Institute (ILSI) has created a database of crop varieties,[60] including historical or unusual varieties that have unusually high or low levels of certain components. It appears that the primary purpose of this database is to provide comparative data that allow industry to argue that the constituents of their GMOs are within the normal range of variation, regardless of how deviant they are from the norm and from the appropriate comparator, which is the relevant non-GM isogenic line grown under the same conditions. EFSA experts use this industry database as the basis of the compositional comparison in GMO risk assessments.[37]

If, on the basis of this "comparative safety assessment", EFSA experts judge the GM crop to be equivalent to the comparator non-GM crops, it is assumed to be as safe.[37,61] Further rigorous tests that could reveal unexpected differences, such as long-term animal feeding trials and environmental stress tests, are not required.[37]

References

1. European Commission. GMOs in a nutshell. 2009. http://ec.europa.eu/food/food/biotechnology/qanda/a1_en.print.htm.
2. Monsanto. Commonly asked questions about the food safety of GMOs. 2013. http://www.monsanto.com/newsviews/Pages/food-safety.aspx.
3. Tokar B. Deficiencies in Federal Regulatory Oversight of Genetically Engineered Crops. Institute for Social Ecology Biotechnology Project; 2006. http://environmentalcommons.org/RegulatoryDeficiencies.html.
4. Freese W, Schubert D. Safety testing and regulation of genetically engineered foods. Biotechnol Genet Eng Rev. 2004:299-324.
5. Michaels D. Doubt Is Their Product: How Industry's Assault on Science Threatens Your Health. Oxford University Press; 2008.
6. Barnes DE, Bero LA. Why review articles on the health effects of passive smoking reach different conclusions. JAMA. 1998;279:1566-1570.
7. Lexchin J, Bero LA, Djulbegovic B, Clark O. Pharmaceutical industry sponsorship and research outcome and quality: systematic review. Br Med J. 2003;326:1167. doi:10.1136/bmj.326.7400.1167.
8. Lexchin J. Those who have the gold make the evidence: How the pharmaceutical industry biases the outcomes of clinical trials of medications. Sci Eng Ethics. 2011. doi:10.1007/s11948-011-9265-3.
9. Bekelman JE, Li Y, Gross CP. Scope and impact of financial conflicts of interest in biomedical research: a systematic review. JAMA. 2003;289:454-465.
10. Huss A, Egger M, Hug K, Huweiler-Müntener K, Röösli M. Source of funding and results of studies of health effects of mobile phone use: Systematic review of experimental studies. Environ Health Perspect. 2007;115:1-4.
11. Diels J, Cunha M, Manaia C, Sabugosa-Madeira B, Silva M. Association of financial or professional conflict of interest to research outcomes on health risks or nutritional assessment studies of genetically modified products. Food Policy. 2011;36:197-203.
12. Kahl L. Memorandum to Dr James Maryanski, FDA Biotechnology Coordinator, about the Federal Register Document, "Statement of Policy: Foods from Genetically Modified Plants." US Food & Drug Administration; 1992. http://www.biointegrity.org.
13. Guest GB. Memorandum to Dr James Maryanski, Biotechnology Coordinator: Regulation of Transgenic Plants – FDA Draft Federal Register Notice on Food Biotechnology. US Department of Health & Human Services; 1992. http://www.biointegrity.org.
14. Matthews EJ. Memorandum to Toxicology Section of the Biotechnology Working Group: "Safety of Whole Food Plants Transformed by Technology Methods." US Food & Drug

Administration; 1991. http://www.biointegrity.org.
15. Shibko SL. Memorandum to James H. Maryanski, Biotechnology Coordinator, CFSAN: Revision of Toxicology Section of the "Statement of Policy: Foods Derived from Genetically Modified Plants." US Food & Drug Administration; 1992. http://www.biointegrity.org.
16. Pribyl LJ. Comments on the March 18, 1992 Version of the Biotechnology Document. US Food & Drug Administration; 1992. http://www.biointegrity.org.
17. Pribyl LJ. Comments on Biotechnology Draft Document, 2/27/92. US Food & Drug Administration; 1992. http://www.biointegrity.org.
18. Sudduth MA. Genetically engineered foods – fears and facts: An interview with FDA's Jim Maryanski. FDA Consum. 1993:11-14.
19. Bittman M. Why aren't GMO foods labeled? New York Times. http://opinionator.blogs.nytimes.com/2011/02/15/why-arent-g-m-o-foods-labeled/. Published February 15, 2011.
20. Nestle M. Food Politics: How the Food Industry Influences Nutrition and Health. Revised 15 October 2007. University of California Press; 2002.
21. US Food and Drug Administration (FDA). Meet Michael R. Taylor, J.D., deputy commissioner for foods. 2013. http://www.fda.gov/AboutFDA/CentersOffices/OfficeofFoods/ucm196721.htm.
22. US Food and Drug Administration (FDA). Biotechnology Consultation Agency Response Letter BNF No. 000001.; 1995. http://www.fda.gov/Food/FoodScienceResearch/Biotechnology/Submissions/ucm161129.htm.
23. Vidal J. WikiLeaks: US targets EU over GM crops. The Guardian. http://www.guardian.co.uk/world/2011/jan/03/wikileaks-us-eu-gm-crops. Published January 3, 2011.
24. Euractiv.com. US lobbied EU to back GM crops: WikiLeaks. http://www.euractiv.com/global-europe/us-lobbied-eu-back-gm-crops-wikileaks-news-500960. Published January 4, 2011.
25. EINNEWS. Wikileaks document pushes genetically modified food for African countries. http://www.einnews.com/pr-news/248883-wikileaks-document-pushes-genetically-modified-food-for-african-countries. Published December 1, 2010.
26. Laskawy T. Wikileaks: State Dept wants intel on African acceptance of GMOs. GRIST. http://www.grist.org/article/2010-11-29-wikileaks-state-dept-wants-intel-on-african-acceptance-of-gmos. Published November 30, 2010.
27. Pollan M. Playing God in the garden. New York Times Magazine. http://www.nytimes.com/1998/10/25/magazine/playing-god-in-the-garden.html. Published October 25, 1998.
28. US Food and Drug Administration (FDA). Statement of policy: Foods derived from new plant varieties. FDA Fed Regist. 1992;57(104):22984.
29. European Food Safety Authority (EFSA). Frequently asked questions on EFSA GMO risk assessment. 2006. http://www.cibpt.org/docs/faq-efsa-gmo-risk-assessment.pdf.
30. US Food and Drug Administration (FDA). CFR - Code of Federal Regulations Title 21, Volume 3 (Revised 1 April 2013): 21CFR170.30. 2013. http://www.accessdata.fda.gov/scripts/cdrh/cfdocs/cfCFR/CFRSearch.cfm?fr=170.30.
31. Maryanski J. Letter from Dr James Maryanski, Biotechnology Coordinator, to Dr Bill Murray, Chairman of the Food Directorate, Canada. Subject: the safety assessment of foods and food ingredients developed through new biotechnology. 1991. http://www.biointegrity.org/FDAdocs/06/view1.html.
32. Royal Society of Canada. Elements of Precaution: Recommendations for the Regulation of Food Biotechnology in Canada. An Expert Panel Report on the Future of Food Biotechnology.; 2001. https://rsc-src.ca/sites/default/files/pdf/GMreportEN.pdf.
33. Hilbeck A, Binimelis R, Defarge N, et al. No scientific consensus on GMO safety. Environ Sci Eur. 2015;27(1):4. doi:10.1186/s12302-014-0034-1.
34. Millstone E, Brunner E, Mayer S. Beyond "substantial equivalence." Nature. 1999;401:525-526. doi:10.1038/44006.
35. Howard CV. GM crops inquiry: Testimony of Prof C. Vyvyan Howard to the Scottish Parliament Health and Community Care Committee, meeting no. 31, 27 November 2002. 2002. http://archive.scottish.parliament.uk/business/committees/historic/health/or-02/he02-3102.htm.
36. Organisation for Economic Cooperation and Development (OECD). Safety Evaluation of Foods Derived by Modern Biotechnology: Concepts and Principles. OECD Publishing; 1993. http://dbtbiosafety.nic.in/guideline/OACD/Concepts_and_Principles_1993.pdf.
37. Then C, Bauer-Panskus A. European Food Safety Authority: A Playing Field for the Biotech Industry. Munich, Germany: Testbiotech; 2010. http://www.testbiotech.de/en/node/431.
38. Levidow L, Murphy J, Carr S. Recasting "substantial equivalence": Transatlantic governance of GM food. Sci Technol Hum Values. 2007;32:26-64.
39. European Parliament and Council. Commission implementing regulation (EU) no. 503/2013 of 3 April 2013 on applications for authorisation of genetically modified food and feed in accordance with Regulation (EC) No 1829/2003 of the European Parliament and of the Council and amending Commission Regulations (EC) No 641/2004 and (EC) No 1981/2006.

Off J Eur Union. 2013. http://eur-lex.europa.eu/LexUriServ/LexUriServ.do?uri=OJ:L:2013:157:0001:0048:EN:PDF.
40. Pusztai A, Bardocz S, Ewen SWB. Genetically modified foods: Potential human health effects. In: D'Mello JPF, ed. Food Safety: Contaminants and Toxins. Wallingford, Oxon: CABI Publishing; 2003:347-372. http://www.leopold.iastate.edu/sites/default/files/events/Chapter16.pdf.
41. Nodari RO, Guerra MP. Implications of transgenics for environmental and agricultural sustainability. Hist Cienc Saude Manguinhos. 2000;7(2):481-491.
42. Zdunczyk Z. In vivo experiments on the safety evaluation of GM components of feeds and foods. J Anim Feed Sci. 2001;10:195-210.
43. Zolla L, Rinalducci S, Antonioli P, Righetti PG. Proteomics as a complementary tool for identifying unintended side effects occurring in transgenic maize seeds as a result of genetic modifications. J Proteome Res. 2008;7:1850-1861. doi:10.1021/pr0705082.
44. Lappé M, Bailey B, Childress C, Setchell KDR. Alterations in clinically important phytoestrogens in genetically modified herbicide-tolerant soybean. J Med Food. 1999;1:241-245.
45. Padgette SR, Taylor NB, Nida DL, et al. The composition of glyphosate-tolerant soybean seeds is equivalent to that of conventional soybeans [Supplementary information from Puerto Rico trials, excluded from the main publication but deposited with the American Society for Information Science, National Auxiliary Publication Service (NAPS) and in the archives of the Journal of Nutrition]. J Nutr. 1996;126(3):702-716.
46. Shewmaker C, Sheehy JA, Daley M, Colburn S, Ke DY. Seed-specific overexpression of phytoene synthase: Increase in carotenoids and other metabolic effects. Plant J. 1999;20:401-412X.
47. Jiao Z, Si XX, Li GK, Zhang ZM, Xu XP. Unintended compositional changes in transgenic rice seeds (Oryza sativa L.) studied by spectral and chromatographic analysis coupled with chemometrics methods. J Agric Food Chem. 2010;58:1746-1754. doi:10.1021/jf902676y.
48. Abdo EM, Barbary OM, Shaltout OE. Chemical analysis of Bt corn "Mon-810: Ajeeb-YG®" and its counterpart non-Bt corn "Ajeeb." IOSR J Appl Chem. 2013;4(1):55-60.
49. Gab-Alla AA, El-Shamei ZS, Shatta AA, Moussa EA, Rayan AM. Morphological and biochemical changes in male rats fed on genetically modified corn (Ajeeb YG). J Am Sci. 2012;8(9):1117-1123.
50. El-Shamei ZS, Gab-Alla AA, Shatta AA, Moussa EA, Rayan AM. Histopathological changes in some organs of male rats fed on genetically modified corn (Ajeeb YG). J Am Sci. 2012;8(10):684-696.
51. Kok EJ, Kuiper HA. Comparative safety assessment for biotech crops. Trends Biotechnol. 2003;21:439-444.
52. European Food Safety Authority (EFSA). Annual Declaration of Interests – Esther Kok.; 2010.
53. International Life Sciences Institute (ILSI). Nutritional and safety assessments of foods and feeds nutritionally improved through biotechnology, prepared by a task force of the ILSI International Food Biotechnology Committee. Compr Rev Food Sci Food Saf. 2004;3:38-104.
54. European Parliament and Council. Directive 2001/18/EC of the European Parliament and of the Council of 12 March 2001 on the deliberate release into the environment of genetically modified organisms and repealing Council Directive 90/220/EEC. Off J Eur Communities. 2001:1-38.
55. Padgette SR, Taylor NB, Nida DL, et al. The composition of glyphosate-tolerant soybean seeds is equivalent to that of conventional soybeans. J Nutr. 1996;126:702-716.
56. Taylor NB, Fuchs RL, MacDonald J, Shariff AR, Padgette SR. Compositional analysis of glyphosate-tolerant soybeans treated with glyphosate. J Agric Food Chem. 1999;47:4469-4473.
57. Hammond B, Dudek R, Lemen J, Nemeth M. Results of a 13 week safety assurance study with rats fed grain from glyphosate tolerant corn. Food Chem Toxicol. 2004;42:1003-1014. doi:10.1016/j.fct.2004.02.013.
58. Hammond B, Lemen J, Dudek R, et al. Results of a 90-day safety assurance study with rats fed grain from corn rootworm-protected corn. Food Chem Toxicol. 2006;44:147-160. doi:10.1016/j.fct.2005.06.008.
59. Hilbeck A, Meier M, Römbke J, Jänsch S, Teichmann H, Tappeser B. Environmental risk assessment of genetically modified plants - concepts and controversies. Environ Sci Eur. 2011;23. doi:10.1186/2190-4715-23-13.
60. International Life Sciences Institute (ILSI). ILSI crop composition database: Version 4. 2011. http://www.cropcomposition.org/query/index.html.
61. European Food Safety Authority (EFSA). Guidance on the submission of applications for authorisation of genetically modified food and feed and genetically modified plants for food or feed uses under Regulation (EC) No 1829/2003. EFSA J. 2011;9:1-27. doi:10.2903/j.efsa.2011.2311.

3 Myth: Independent studies confirm that GM foods and crops are safe

Truth: Independent research on GM foods is difficult or impossible to carry out, but many existing studies have found problems

Myth at a glance

In-depth food safety studies on GM crops and foods carried out by scientists independent of the GMO industry are rare. They are hampered by the lack of funding for genuinely independent research and the difficulty of accessing GM seeds and the non-GM parent varieties.

Scientists who have managed to carry out such research and have found risks from the genetically modified organism (GMO) tested have suffered persecution. Some have paid with their careers and funding.

Claims that the climate for independent researchers has improved in recent years remain unproven. No examples of agreements between researchers and GMO seed companies are in the public domain, so the public cannot see what limitations are imposed on researchers.

The pro-GMO compilers of a database of studies on GMO risks claimed that half of the studies in the database were independent of industry and that overall these studies show GMOs are safe.[1] However, these claims have been exposed as flawed. Monsanto was shown to be the first or the second (after the US government) most common funder of studies in the database and some studies classified by the compilers as independent were in fact funded by Monsanto. The US government promotes GM crops (see Chapter 2) and thus is not independent or impartial.[2]

Genuinely independent studies on GM foods and crops are rare, because such research is not supported financially. In some cases, research may appear independent but is carried out by researchers at universities or institutions that rely on GMO industry funding. Such interests often are not declared on research papers, where authors may be listed only under their public affiliations.

Publicly funded universities, institutions, and scientists can no longer be assumed to be independent. GMO companies have representatives on university boards and fund research, buildings, and departments.[3] Monsanto has donated at least a million dollars to the University of Florida Foundation.[4,5] Many US land grant universities that do crop research are beholden to Monsanto.[6] Academic scientists own GMO patents, are involved in spin-off companies that develop GM crops,[7] and are awarded lucrative consultancy contracts by industry.[3] In the UK, the public Rothamsted Research Institute counts Monsanto as a collaborator.[8] Universities have become businesses and scientists have become entrepreneurs and salespeople. All this should be borne in mind when evaluating claims about GM technology by so-called independent scientists.

The GMO industry restricts access to seeds for independent research

Scientists who want to find out if a GM crop is safe to eat or harms the environment need access to seeds of the GM variety and the non-GM parent (isogenic) variety it was developed from, grown under the same conditions. But researchers are often denied access to these seeds.[9,10] Even if permission to carry out research is given, GM companies typically retain the right to block publication.[9,10] An editorial in Scientific American reported, "Only studies that the seed companies have approved ever see the light of a peer-reviewed journal. In a number of cases, experiments that had the implicit go-ahead from the seed company were later blocked from publication because the results were not flattering."[11]

Scientists protest

In 2009, 26 scientists made a formal complaint to the US Environmental Protection Agency. They wrote, "No truly independent research can be legally conducted on many critical questions involving these crops."[12]

In response to the furore that followed, a new agreement for researchers on GM crops was reached between US Department of Agriculture (USDA) scientists and Monsanto in 2010. However, this agreement is still restrictive. It only applies to agronomic, not food safety research, and only to USDA scientists.[13] Given that the US government has a policy of supporting the GMO industry,[14] perhaps Monsanto does not see USDA scientists as a threat.

Is the problem of access to research materials solved?

In 2013 the food writer Nathanael Johnson concluded that before 2009, some researchers had trouble investigating GMOs due to patent restrictions, but that now the problem was "largely fixed", due to research agreements being reached between GMO companies and universities.[15]

The independent scientist Dr Judy Carman, who has researched the effects of feeding GMOs to pigs,[16] is skeptical of this view. Carman said: "GM crops are under patent protection. This means that you cannot go to a seed merchant to buy GM seeds to test. If you do, you will be presented with a legal contract (a technology user agreement[17]) to sign that states that you will not do any research on the seeds and you will not give them to anyone else to do research on, either. We tried this approach and I've seen the agreements.[18]

"The way US schools get access to GM crops is by signing commercial in-confidence agreements that you and I cannot see. So we cannot see the conditions placed upon the researchers and the institutions involved.

"Also, GM companies will only reach these agreements with US schools that they approve of – typically those that work in partnership with the GM company to make GM crops where both can benefit financially."[18]

Carman had approached three GM companies directly to source GM seeds and the non-GM isogenic varieties for her toxicity study on pigs.[16] One company didn't reply. Another wanted the details of her study first. Monsanto sent her a legal document to sign stating that she would give the company the results of the study before publication. Carman said: "We would have been legally bound to do that whether they gave us seeds or not. No sensible scientist would agree to such conditions, and we didn't."[19]

Finally Carman used non-isogenic varieties for the control diets in her pig study, pointing out that government regulators had declared the GM varieties substantially equivalent to non-GM varieties, and so there should be no differences between the GM- and non-GM-fed animals. The regulators were wrong, however, because Carman's team found toxic effects, including a higher incidence of severe stomach inflammation and heavier uteri, in the GM-fed pigs.[16]

Another researcher finds problems accessing materials

Professor Gilles-Eric Séralini, a molecular biologist at the University of Caen, France, also had difficulty accessing crops for his long-term rat feeding study on a GMO and its associated pesticide, Roundup.[20] Séralini approached farmers in Spain, Romania, and the US, to grow the crops, but none of them would risk breaching their technology agreement with Monsanto, which forbids the use of GM seed for research. Eventually a farm school in Canada agreed to grow the crops on condition that it was not named, as Séralini said, "out of fear of reprisal from the seed suppliers".[20]

Claims that access problems are solved remain unproven

> *"Unfortunately, it is impossible to verify that genetically modified crops perform as advertised. That is because agritech companies have given themselves veto power over the work of independent researchers... Only studies that the seed companies have approved ever see the light of a peer-reviewed journal. In a number of cases, experiments that had the implicit go-ahead from the seed company were later blocked from publication because the results were not flattering... It would be chilling enough if any other type of company were able to prevent independent researchers from testing its wares and reporting what they find... But when scientists are prevented from examining the raw ingredients in our nation's food supply or from testing the plant material that covers a large portion of the country's agricultural land, the restrictions on free inquiry become dangerous."*
>
> – Editorial, Scientific American[11]

The examples of Carman and Séralini date from some years ago. But claims that the situation has improved in recent years remain unproven, especially for researchers who insist on maintaining control over their research and findings. Those who claim that the problem of access to research materials has been solved must prove it by publishing the agreements that researchers have signed with GMO companies. The public would then be able to see for themselves how much freedom the researchers have to investigate critical biosafety questions around GMOs and publish the research without restriction.

Researchers who publish studies that find harm from GM crops are attacked

Sometimes independent researchers succeed in carrying out critical research on GMOs. But their problems are not over – in fact, they are just beginning.

This is because the GMO lobby uses a range of public relations strategies to discredit the research and the researchers themselves.[21]

In some cases pro-GM scientists have bullied the journal editor to try to persuade them not to publish the study. If it is published, they criticize it as "bad science", identifying any flaw or limitation (which all studies have) and claiming that this invalidates all the findings.

The proper way to move scientific knowledge forward is to carry out a further study to clarify, confirm, or refute the findings of the initial study. This does not happen in the area of GMO research. Instead the critics "shout down" the study on the basis of claims that are spurious or not scientifically validated. The following are just two examples among many.

Gilles-Eric Séralini

In 2012 the French researcher Professor Gilles-Eric Séralini and his research team were subjected to a vicious smear campaign after he and his research colleagues published a study showing that rats fed over a two-year period with Monsanto's GM maize NK603 and very low levels of the Roundup herbicide it is engineered to tolerate suffered severe organ damage. They also showed a trend of increased rates of tumours and premature death.[22]

Many of Séralini's attackers had links with the GM industry or with organizations with vested interests in the public acceptance of GM technology. These links and vested interests went undeclared in media articles that quoted them.[23,24]

More criticism of the study came from government agencies that had previously given opinions that this or other GM foods were safe and were therefore not independent authorities – such as the European Food Safety Authority (EFSA).[25,26,27] In an additional twist, a report found that over half of EFSA's experts have conflicts of interest with the industries they regulate.[28]

The smear campaign against the study quickly focused on trying to persuade the editor-in-chief of the journal that initially published it, A. Wallace Hayes, to retract it.[23] Over a year after the study had been published, Hayes retracted it.[29] The reasons he gave for the retraction – the supposed "inconclusive" nature of some of the findings – were scientifically unjustified and unprecedented in scientific publishing.[30]

The retraction was condemned by hundreds of scientists worldwide in a series of articles, letters and petitions.[31,32,33,34,35,36] Many scientists derided the notion that a scientific study should produce "conclusive" results.[32,36,37] Professor Jack Heinemann of the University of Canterbury, New Zealand wrote that if Hayes's criterion of conclusiveness were universally applied, then among the studies that would have to be retracted were two pioneering papers by Nobel Prize winners James Watson and Francis Crick, describing the structure of DNA and how it might replicate. At the time of publication the findings were inconclusive.[37]

David Schubert, a professor at the Salk Institute for Biological Studies in the USA, wrote, "As a scientist, I can assure you that if this [inconclusiveness] were a valid reason for retracting a publication, a large fraction of the scientific literature would not exist."[36]

In a later statement attempting to justify the retraction, Hayes claimed that the paper was an example of "honest error", since it claimed a "definitive link between GMO and cancer".[38] However, Séralini's paper nowhere claimed that GM maize NK603 and/or Roundup have a link to cancer, let alone a definitive link. The word "cancer" is not used in the paper and the authors stated in their introduction that the study was not designed as a carcinogenicity study, but a chronic toxicity study.[22] The tumour observations came as a surprise.

Séralini's study was republished in 2014 by another journal.[39] It remains in the literature as a citable study.

Manuela Malatesta

The Italian researcher Manuela Malatesta and her team carried out studies which found that Monsanto's GM soy disturbed the functioning of the liver, pancreas and testes of mice.[40,41,42,43] After Malatesta published her papers, she was forced out of her job at the university where she had worked for ten years and could not obtain funding to follow up her research. She said, "Research on GMOs is now taboo. You can't find money for it... People don't want to find answers to troubling questions. It's the result of widespread fear of Monsanto and GMOs in general."[44]

References

1. Von Mogel KH. Announcing the launch of the GENERA beta test. Biology Fortified, Inc. http://www.biofortified.org/2014/08/announcing-the-launch-of-the-genera-beta-test/. Published August 25, 2014.
2. Schwab T. Pro-GMO database: Monsanto is most common funder of GMO research. Food & Water Watch. http://www.foodandwaterwatch.org/blogs/pro-gmo-database-monsanto-most-common-funder-gmo-research/. Published September 16, 2014.
3. Food & Water Watch. Public Research, Private Gain: Corporate Influence over University Agricultural Research. Washington, DC: Food & Water Watch; 2012. http://documents.foodandwaterwatch.org/doc/PublicResearchPrivateGain.pdf.
4. University of Florida Foundation. 2013/2014 Honor Roll of Donors: President's Council – Gold Level. 2014. https://www.uff.ufl.edu/honorroll/Organizations.asp?LN=Gold.
5. University of Florida Foundation. 2013/2014 Honor Roll of Donors. 2014. https://www.uff.ufl.edu/HonorRoll/Default.asp.
6. Iowa State University Foundation. Monsanto. Iowa State Univ. 2013. http://www.foundation.

iastate.edu/s/1463/index.aspx?pgid=1027&gid=1&cid=3248.
7. GeneWatch UK. Commercial interests. GeneWatch UK. 2015. http://www.genewatch.org/sub-568236.
8. Rothamsted Research. Collaborators: Monsanto UK Ltd. Rothamsted Res. 2015. http://www.rothamsted.ac.uk/collaborators/monsanto-uk-ltd.
9. Waltz E. Battlefield. Nature. 2009;461:27-32. doi:10.1038/461027a.
10. Waltz E. Under wraps – Are the crop industry's strong-arm tactics and close-fisted attitude to sharing seeds holding back independent research and undermining public acceptance of transgenic crops? Nat Biotechnol. 2009;27(10):880-882. doi:10.1038/nbt1009-880.
11. Scientific American. Do seed companies control GM crop research?http://www.scientificamerican.com/article.cfm?id=do-seed-companies-control-gm-crop-research. Published July 20, 2009.
12. Pollack A. Crop scientists say biotechnology seed companies are thwarting research. New York Times. http://www.nytimes.com/2009/02/20/business/20crop.html. Published February 19, 2009.
13. Waltz E. Monsanto relaxes restrictions on sharing seeds for research. Nat Biotechnol. 2010;28:996. doi:10.1038/nbt1010-996c.
14. Sudduth MA. Genetically engineered foods – fears and facts: An interview with FDA's Jim Maryanski. FDA Consum. 1993:11-14.
15. Johnson N. Food for bots: Distinguishing the novel from the knee-jerk in the GMO debate. Grist. 2013. http://grist.org/food/dodging-argument-bot-crossfire-to-revisit-some-gm-research-controversies/.
16. Carman JA, Vlieger HR, Ver Steeg LJ, et al. A long-term toxicology study on pigs fed a combined genetically modified (GM) soy and GM maize diet. J Org Syst. 2013;8:38-54.
17. Monsanto. 2011 Monsanto technology/stewardship agreement (limited use license). 2010. http://thefarmerslife.files.wordpress.com/2012/02/scan_doc0004.pdf.
18. GMOJudyCarman. How easy is it for researchers to access the materials for GM biosafety research?http://gmojudycarman.org/how-easy-is-it-for-researchers-to-access-the-materials-for-gm-biosafety-research/. Published September 1, 2013.
19. Carman J. Accessing GM seeds and non-GM isolines for GMO safety research [personal email communication]. 2014.
20. Séralini GE. Tous Cobayes!. Paris, France: Flammarion; 2012.
21. Lotter D. The genetic engineering of food and the failure of science – Part 2: Academic capitalism and the loss of scientific integrity. Int Jrnl Soc Agr Food. 2008;16:50-68.
22. Séralini GE, Clair E, Mesnage R, et al. [RETRACTED:] Long term toxicity of a Roundup herbicide and a Roundup-tolerant genetically modified maize. Food Chem Toxicol. 2012;50:4221-4231.
23. Matthews J. Smelling a corporate rat. Spinwatch. http://www.spinwatch.org/index.php/issues/science/item/164-smelling-a-corporate-rat. Published December 11, 2012.
24. Sourice B. OGM: La guerre secrète pour décrédibiliser l'étude Séralini [The covert war to discredit Séralini's study]. Rue 89/Le Nouvel Observateur. http://blogs.rue89.nouvelobs.com/de-interet-conflit/2012/11/12/ogm-la-guerre-secrete-pour-decredibiliser-letude-seralini-228894. Published November 12, 2012.
25. European Food Safety Authority (EFSA). Review of the Séralini et al. (2012) publication on a 2-year rodent feeding study with glyphosate formulations and GM maize NK603 as published online on 19 September 2012 in Food and Chemical Toxicology. EFSA J. 2012;10:2910.
26. European Food Safety Authority (EFSA). Final review of the Séralini et al. (2012a) publication on a 2-year rodent feeding study with glyphosate formulations and GM maize NK603 as published online on 19 September 2012 in Food and Chemical Toxicology. EFSA J. 2012;10:2986.
27. European Food Safety Authority (EFSA). Opinion of the Scientific Panel on Genetically Modified Organisms on a request from the Commission related to the safety of foods and food ingredients derived from herbicide-tolerant genetically modified maize NK603, for which a request for placing on the market was submitted under Article 4 of the Novel Food Regulation (EC) No 258/97 by Monsanto (QUESTION NO EFSA-Q-2003-002): Opinion adopted on 25 November 2003. EFSA J. 2003;2003(9):1-14.
28. Corporate Europe Observatory. Unhappy Meal: The European Food Safety Authority's Independence Problem.; 2013. http://corporateeurope.org/sites/default/files/attachments/unhappy_meal_report_23_10_2013.pdf.
29. Elsevier. Elsevier announces article retraction from Journal Food and Chemical Toxicology. 2013. http://www.elsevier.com/about/press-releases/research-and-journals/elsevier-announces-article-retraction-from-journal-food-and-chemical-toxicology#sthash.VfY74Y24.dpuf.
30. Hayes AW. Letter to Professor GE Séralini. 2013. http://www.gmwatch.org/files/Letter_AWHayes_GES.pdf.

31. Portier CJ, Goldman LR, Goldstein BD. Inconclusive findings: Now you see them, now you don't! Environ Health Perspect. 2014;122(2).
32. EndScienceCensorship.org. Statement: Journal retraction of Séralini GMO study is invalid and an attack on scientific integrity. 2014. http://www.endsciencecensorship.org/en/page/Statement#.UwUSP14vFY4.
33. Institute of Science in Society. Open letter on retraction and pledge to boycott Elsevier. 2013. http://www.i-sis.org.uk/Open_letter_to_FCT_and_Elsevier.php#form.
34. AFP. Mexican scientists criticise journal's retraction of study on GMO. terra.cl. http://bit.ly/1jVl1HZ ; English translation available at: http://gmwatch.org/index.php/news/archive/2013/15225. Published December 18, 2013.
35. European Network of Scientists for Social and Environmental Responsibility (ENSSER). Journal's Retraction of Rat Feeding Paper Is a Travesty of Science and Looks like a Bow to Industry: ENSSER Comments on the Retraction of the Séralini et Al. 2012 Study. Berlin, Germany; 2013. http://bit.ly/1cytNa4.
36. Schubert D. Science study controversy impacts world health. U-T San Diego. http://www.utsandiego.com/news/2014/jan/08/science-food-health/. Published January 8, 2014.
37. Heinemann J. Let's give the scientific literature a good clean up. Biosafetycooperative.newsvine.com. 2013. http://bit.ly/1aeULiB.
38. Hayes AW. Food and Chemical Toxicology editor-in-chief, A. Wallace Hayes, publishes response to Letters to the Editors. 2013. http://www.elsevier.com/about/press-releases/research-and-journals/food-and-chemical-toxicology-editor-in-chief,-a.-wallace-hayes,-publishes-response-to-letters-to-the-editors#sthash.tTW2LCGq.dpuf.
39. Séralini G-E, Clair E, Mesnage R, et al. Republished study: long-term toxicity of a Roundup herbicide and a Roundup-tolerant genetically modified maize. Environ Sci Eur. 2014;26(14). doi:10.1186/s12302-014-0014-5.
40. Vecchio L, Cisterna B, Malatesta M, Martin TE, Biggiogera M. Ultrastructural analysis of testes from mice fed on genetically modified soybean. Eur J Histochem. 2004;48:448-454.
41. Malatesta M, Caporaloni C, Gavaudan S, et al. Ultrastructural morphometrical and immunocytochemical analyses of hepatocyte nuclei from mice fed on genetically modified soybean. Cell Struct Funct. 2002;27:173-180.
42. Malatesta M, Caporaloni C, Rossi L, et al. Ultrastructural analysis of pancreatic acinar cells from mice fed on genetically modified soybean. J Anat. 2002;201:409-415.
43. Malatesta M, Biggiogera M, Manuali E, Rocchi MBL, Baldelli B, Gazzanelli G. Fine structural analyses of pancreatic acinar cell nuclei from mice fed on genetically modified soybean. Eur J Histochem. 2003;47:385-388.
44. Robin M-M. The World According to Monsanto: Pollution, Corruption, and the Control of the World's Food Supply. The New Press; 2013.

4 **Myth:** GM foods are safe to eat

Truth: GM crops have toxic and allergenic effects on laboratory and farm animals

Myth at a glance

It is often claimed that there is no evidence of dangers to health from GM crops and foods. But this is false. Peer-reviewed studies have found potential signs of toxicity and actual toxic and allergenic effects on the health of laboratory and farm animals fed GMOs.

Most animal feeding studies on GMOs are only short-term or medium-term in length. What is needed are long-term and multi-generational studies on GMOs to see if the changes found in short- and medium-term studies, which are often suggestive of harmful health effects, develop into serious disease, premature death, or reproductive or developmental effects. Such long-term studies are not required by regulators anywhere in the world.

Industry and regulators often dismiss findings of toxicity in animal feeding trials on GMOs by claiming they are not "biologically significant/ relevant". However, these terms have never been properly defined in the context of animal feeding trials with GMOs and are scientifically meaningless.

Claims that "EU research" shows GMOs are safe are false. The EU research project in question was not designed to look at the safety of any particular GMO and did not test any commercialized GMO. Only a small handful of short animal feeding studies were published as a result of the project, and all found problems with the GMO tested.

Claims that Americans have eaten millions of GM meals with no ill effects are unscientific and dishonest. No human epidemiological studies have been done to establish whether GM foods may be affecting Americans' health. Also, GM foods are not labelled in that nation, so there is no way of tracing any effects. What is known is that the health of Americans has got markedly worse since GM foods were introduced. The cause is unknown but GM foods and/or their associated pesticides cannot be ruled out.

Feeding studies on laboratory animals and farm livestock have found that some GM crops, including those already commercialized, have toxic or allergenic effects. There are three major possible sources of adverse health effects from GM foods:

- *Process:* The GM transformation process causes mutations. Depending on where in the genome these occur, these can disrupt or alter gene structure, disturb normal gene regulatory processes, or cause effects at other levels of biological structure and function. These effects can result in unintended changes in food composition, including new toxins or allergens and/or altered nutritional value.
- *Gene product:*
  - The GM gene product – for example, Bt insecticidal toxin – may be toxic or allergenic in itself.
  - If the GM gene product is an enzyme, for example, bacterial EPSPS-CP4 in crops engineered to tolerate application of glyphosate-based herbicides, it may be altered in function in its new GM plant host. It may then catalyze, in addition to the intended reaction, other reactions that produce compounds that are toxic or allergenic, or that alter the nutritional value of the food. The latter may include the production of anti-nutrients.
  - The GM gene product may influence the regulation of molecular processes other than those intended, resulting in production of compounds that are allergenic or toxic. Alternatively it may reduce the production of compounds that are important for the health of the crop or for its nutritional value. One example of such GM gene products is a protein that controls the function of other genes ("transcription factor"), as in the case of the snapdragon genes engineered into tomatoes to produce purple anthocyanin antioxidants.[1,2] A second example of such a GM gene product is small hairpin RNAs (shRNAs), a type of double-stranded RNA molecule involved in reducing the expression of genes. These shRNAs are intended either to impose control on host plant genes or to act as insecticides. Regulators have not assessed the risks of shRNAs adequately,[3] even though they are known to have off-target effects, such as reducing the expression of non-target genes. Crucially, similar molecules in plants known as miRNAs have been found to affect gene expression and the functioning of important bodily processes in mice that ate the plants.[4] Thus the shRNAs engineered into GM plants may have off-target effects – not only in the crop, but also in the consumer, with potentially negative health outcomes.

- *Cultivation practices:* Changes in farming practices linked to the use of a GMO may result in toxic residues – for example, higher levels of crop contamination with the herbicide Roundup are an inevitable result of using GM Roundup Ready crops.[5]

Studies suggest that problems are arising from all three sources, as detailed below.

Toxic and allergenic effects found in GM-fed animals

Altered blood biochemistry, multiple organ damage, and potential effects on male fertility

Rats fed the GM Bt maize MON810: Ajeeb YG (a variety developed by Monsanto for the Egyptian market) for 45 and 91 days showed differences in organ and body weights and in blood biochemistry, compared with rats fed the non-GM parent variety grown side-by-side in the same conditions. The authors noted that the changes could indicate "potential adverse health/ toxic effects", which needed further investigation.[6]

Histopathological investigations by the same researchers found toxic effects in multiple organs in rats fed the GM Bt maize for 91 days. Effects included abnormalities and fatty degeneration of liver cells, congestion of blood vessels in kidneys, and excessive growth and necrosis (death) of intestinal structures called villi. Examination of the testes revealed necrosis and desquamation (shedding) of the spermatogonial cells that are the foundation of sperm cells and thus of male fertility.[7]

Stomach lesions and unexplained mortality

Rats fed GM tomatoes over a 28-day period developed stomach lesions (sores or ulcers).[8,9] There was unexplained high mortality in GM-fed rats: seven out of 40 rats fed GM tomatoes died within two weeks of the start of the experiment.[10]

A "repeat" study performed by Calgene, the company that developed the tomato, found lesions in non-GM-fed as well as GM-fed animals. However, the study was not in reality a repeat but used tomatoes that had been prepared in a different way, which could affect the results, as noted by Fred Hines, an FDA pathologist. Hines concluded that Calgene had not provided enough data to justify its claim that the lesions seen across all the experiments were "incidental" and not due to the GM tomato.[8]

Immune response and allergic reaction

Mice fed GM peas engineered with an insecticidal protein (alpha-amylase inhibitor) from beans showed a strong, sustained immune reaction against the GM protein. Mice developed antibodies against the GM protein and an allergic-type inflammation response (delayed hypersensitivity reaction). Also, the mice fed on GM peas developed an immune reaction to chicken egg white protein. The mice did not show immune or allergic-type inflammation reactions to either non-GM beans naturally containing the insecticide protein, to egg white protein fed with the natural protein from the beans, or to egg white protein fed on its own.[11]

The findings showed that the GM insecticidal protein acted as a sensitizer, making the mice susceptible to developing immune reactions and allergies to normally non-allergenic foods.[11] This is called immunological cross-priming.

This study is often claimed by GMO proponents to show the effectiveness of the regulatory procedure in preventing allergenic GM foods reaching the market. This is false, since such studies are not required under any regulatory regime.

One of the researchers on the original study subsequently co-authored a second study,[12] which he claimed[13,14] resolved concerns raised by the first study.[11] But this claim is unfounded, as the two studies used different methodologies to evaluate immune reactions.

In the first study, the food was fed to the mice intragastrically (into the stomach), an approximation of human dietary exposure; then the mice were tested for allergic reactions.[11] In the second study, the GM and non-GM test proteins were first injected into the abdomen of the mice (intraperitoneal immunization) or introduced into their noses (intranasal immunization). After this procedure the mice were fed intragastrically with GM peas and non-GM beans containing the test proteins. Then the mice were tested for allergic sensitization. The result: both GM peas and non-GM beans were found to be equally allergenic.[12]

However, the allergic reactions in the second study to both the GM and non-GM test proteins are not surprising, because the mice had already been immunologically sensitized to these products by intraperitoneal and intranasal immunization, prior to feeding.

Therefore the second study[12] does not disprove the allergenic potential of the protein in the GM peas found in the first study.[11] Instead the second study shows that it is possible to induce an allergic response to either GM peas or non-GM beans by pre-immunizing the mice to the proteins in a way that is different from the usual way an animal or human is exposed to a food.

Immune disturbances

Young and old mice fed GM Bt maize for periods of 30 and 90 days showed a marked disturbance in immune system cells and in biochemical activity. Bt maize consumption was also linked to an increase in serum cytokines (protein molecules that can influence the immune response), an effect associated with allergic and inflammatory responses.[15]

A study in rats fed GM Bt rice for 28 or 90 days found a Bt-specific immune response in the non-GM-fed control group as well as the GM-fed groups. The researchers concluded that the immune response in the control animals was due to their inhaling particles of the powdered Bt toxin-containing feed consumed by the GM-fed group. They recommended that for future tests involving Bt crops, GM-fed and control groups should be kept separate.[16] This indicates that animals can be sensitive to small amounts of GM proteins, so even low levels of contamination of conventional crops with GMOs could be harmful to health.

Enlarged lymph nodes and immune disturbances

Mice fed for five consecutive generations with GM herbicide-tolerant triticale (a wheat/rye hybrid) showed enlarged lymph nodes and increased white blood cells, as well as a significant decrease in the percentage of T lymphocytes in the spleen and lymph nodes and of B lymphocytes in lymph nodes and blood, in comparison with controls fed with non-GM triticale.[17] T and B lymphocytes are white blood cells involved in immunity.

Reduced nutritional value and immunological protection of colostrum and milk

The colostrum and milk of goats fed a GM soy diet contained a significantly lower percentage of protein and fat, compared with animals fed on a non-GM soy diet. The colostrum of GM soy-fed animals contained readily detectable GM transgene (EPSPS) fragments. The offspring (kids) of the GM and non-GM control soy-fed goats differed significantly in numerous measurements. These included:

- A significantly lower weight at birth and at the time of slaughter at 30 days of age in the GM-fed group.
- Greater height at the withers and chest width in the non-GM soy-fed control group.
- Antibody (IgG) levels in colostrum and kids' serum was lower in the GM-fed group.

Thus the colostrum and milk of the GM soy-fed goats was not as nutritious and afforded a lower degree of immunological protection than that from the non-GM soy-fed animals.[18]

Disturbed liver, pancreas and testes function

Mice fed GM soy showed disturbed liver, pancreas and testes function. The researchers found abnormally formed nuclei and nucleoli (structures within the nuclei) in liver cells, which indicates increased metabolism and potentially altered patterns of gene expression.[19,20,21]

Liver ageing

Mice fed GM soy over a long-term (24-month) period showed changes in the expression of proteins relating to hepatocyte (liver cell) metabolism, stress response, and calcium signalling, indicating more acute signs of ageing in the liver, compared with the control group fed non-GM soy.[22]

Liver and kidney damage and hormonal disruption

Severe damage to the liver, kidney, and pituitary gland was found in rats fed a commercialized GM maize, called NK603, and tiny amounts of the Roundup herbicide it is grown with over a long-term period of two years. Additional unexpected observations (which need to be confirmed in experiments with larger numbers of animals) were increased rates of large tumours and mortality in most treatment groups of rats. Similar toxic effects were found from GM maize that had not been treated with Roundup, GM maize sprayed with Roundup, and Roundup on its own.[23]

Those who claim that this study, performed by Professor Gilles-Eric Séralini and colleagues, was flawed must apply the same critical standards to the short Monsanto study on the same GM maize. The Monsanto study concluded the maize was safe and was used to justify regulatory approvals. Monsanto used the same strain of rat and analyzed the same number of rats that Séralini's team used, but followed the rats for the much shorter period of 90 days. The Monsanto authors found statistically significant changes in the GM-fed rats, particularly in liver and kidney parameters, but claimed they were not biologically meaningful.[24]

Séralini and colleagues extended the study period to two years and found that the statistically significant changes in liver and kidney parameters found in Monsanto's short study were indeed biologically meaningful, as they escalated over time into increased rates of serious organ damage and hormonal disturbances. In addition, an increase in tumour incidence was noted in females in the Roundup treatment groups, particularly at the lowest dose, where it was statistically significant. This only became evident in the second year of life.[23] Overall, the study showed that the 90-day studies routinely done in support of GMO regulatory approvals are too short to see long-term health effects like serious organ damage, tumours, and premature death.

No other studies have been performed on the long-term safety of NK603 maize, so Séralini's results have not been challenged by hard data.

The study was widely attacked by pro-GMO critics, many of whom had undisclosed conflicts of interest with the GMO industry and/or GM technology.[25] After months of sustained pressure, the editor-in-chief of the journal that first published the study retracted it[26,27] – a decision that was condemned by hundreds of scientists worldwide for not being based on scientific reasons.[28,29,30] The study was subsequently republished by another journal[23] after another round of peer review.[31]

A peer-reviewed analysis compared Séralini's paper to Monsanto's study on the same GM maize, according to criteria used by the European Food Safety Authority (EFSA) to reject Séralini's paper alone. The authors concluded that EFSA had applied unscientific double standards in rejecting Séralini's paper while accepting Monsanto's study at face value.[32]

Higher density of uterine lining

Female rats fed GM soy for 15 months showed significant changes in the uterus and ovaries compared with rats fed organic non-GM soy or a non-soy diet. The number of corpora lutea, structures that secrete sex hormones and are involved in establishing and maintaining pregnancy, was increased only in the GM soy rats compared with the organic soy-fed and non-soy-fed rats. The density of the epithelium (lining of the uterus) was higher in the GM soy-fed group than the other groups, meaning that there were more cells than normal.

Certain effects on the female reproductive system were found with organic soy as well as GM soy when compared with the non-soy diet, leading the authors to conclude that there was a need for further investigation into the effects of soy-based diets (whether GM or non-GM) on reproductive health.[33]

Severe stomach inflammation and heavier uteri

A feeding study in pigs fed a mixed diet containing GMO soy and maize over an average commercial lifespan of 22.7 weeks found that the GM-fed pigs had more severe stomach inflammation than pigs fed an equivalent non-GM diet and 25% heavier uteri, which could be an indicator of pathology.[34] GM-fed pigs had a higher rate of severe stomach inflammation, 32% for GM-fed pigs compared to 12% for non-GM-fed. The severe stomach inflammation was worse in GM-fed males compared with non-GM fed males by a factor of

4.0, and in GM-fed females compared with non-GM fed females by a factor of 2.2.

GMO proponents claimed that non-GM-fed pigs had more cases of mild and moderate inflammation than GM-fed pigs and that therefore the GM diet had a protective effect.[35] However, this claim collapses when it is considered that many GM-fed pigs were moved up from the "mild" and "moderate" categories into the "severe" inflammation category, leaving fewer pigs in the "mild" and "moderate" categories.

Liver and kidney toxicity

A review of 19 studies (including the GMO industry's own studies submitted in support of regulatory authorization of GM crops) on mammals fed with commercialized GM soy and maize found consistent signs of toxicity in the liver and kidneys. Such effects may mark the onset of chronic disease, but longer-term studies are required to assess this potential more thoroughly. Such long-term feeding trials on GMOs are not required by regulators anywhere in the world.[36]

In a separate study, the same research group, led by Professor Gilles-Eric Séralini at the University of Caen, France, re-analyzed Monsanto's own rat feeding trial data, submitted to obtain approval in Europe for three commercialized GM Bt maize varieties. Séralini's team concluded that the maize varieties caused signs of toxicity in liver and kidneys.[37] The data suggest that approval of these GM maize varieties should be withdrawn from the market because they are not substantially equivalent to non-GM maize and may be toxic.

Toxic effects on liver and kidneys and altered blood biochemistry

Rats fed GM Bt maize over three generations showed damage to liver and kidneys and alterations in blood biochemistry.[38]

Enlarged liver

In an experiment conducted by Monsanto scientists, rats fed the company's GM glyphosate-tolerant oilseed rape (canola) over four weeks developed enlarged livers, often a sign of toxicity. Monsanto followed up with another experiment, this time comparing the GM canola with a range of eight different canola varieties, thus widening the range of variation and obscuring any effects of feeding the GM canola. This allowed Monsanto to conclude that the canola was as safe as other canola varieties.[39]

Disturbances in digestive system and changes to liver and pancreas

Female sheep fed Bt GM maize over three generations showed disturbances in the functioning of the digestive system, while their lambs showed cellular changes in the liver and pancreas.[40]

Excessive growth in the lining of the gut

Rats fed GM potatoes for only ten days showed excessive growth of the lining of the gut similar to a pre-cancerous condition, as well as toxic effects in multiple organ systems.[41,42] The GM potatoes were engineered to express a protein (GNA) from snowdrops, which has insecticidal properties. In its natural non-GMO form, the GNA protein is harmless to mammals. As animals fed non-GM potatoes spiked with GNA protein did not suffer adverse health effects, this study clearly implies that novel toxicity arising from the mutagenic effect of the GM transformation process was the most likely cause of the observed problems in the GM potato-fed group.

Intestinal abnormalities

Mice fed a diet of GM Bt potatoes or non-GM potatoes spiked with natural Bt toxin protein isolated from bacteria over two weeks showed abnormalities in the cells and structures of the small intestine, compared with a control group of mice fed non-GM potatoes. The abnormalities were more marked in the Bt toxin-fed group.[43]

This study shows that the GM Bt potatoes caused mild damage to the intestines. It also shows that Bt toxin protein is not harmlessly broken down in digestion, as regulatory agencies claim,[44] but survives in a functionally active form in the small intestine and can cause damage to that organ.[43]

Altered blood biochemistry and gut bacteria, and immune response

GM rice was engineered with a gene from snowdrops to express the GNA insecticidal protein. The GM rice grain was statistically significantly different in composition compared to the non-GM parent rice with regard to many substances such as sugars, protein, starch, minerals, and certain fats and amino acids. This implies that the GM transformation process had brought about numerous unintended changes in gene function, causing alterations in plant biochemistry.

"The argument advanced... for the safety of GM food is false... Yes, the DNA of all living organisms is made up of just four nucleosides, and yes, virtually all proteins are made up from just 20 amino acids. But this does not imply that everything containing these basic building blocks is without risk to human beings. The same units, arranged in different ways, are contained in the smallpox virus, bubonic plague and influenza, deadly nightshade and other poisonous plants, creatures such as poisonous jellyfish, scorpions, deadly snakes, sharks – and people who talk absolute nonsense."

– G. D. W. Smith, Fellow of the Royal Society, professor of materials, Oxford University, UK[47]

"Most studies with GM foods indicate that they may cause hepatic, pancreatic, renal, and reproductive effects and may alter haematological [blood], biochemical, and immunologic parameters, the significance of which remains to be solved with chronic toxicity studies."

– Artemis Dona, Department of Forensic Medicine and Toxicology, University of Athens Medical School, Greece, and Ioannis S. Arvanitoyannis, University of Thessaly School of Agricultural Sciences, Greece[48]

Rats fed for 90 days with the GM rice had a higher water intake as compared with the control group fed the non-GM isogenic line of rice. The GM-fed rats showed differences in blood biochemistry, with some measurements indicating potential damage to the liver. The population of gut bacteria in the GM rice fed rats was altered, implying dysbiosis (microbial imbalance), which could lead to severe ill health in the long term. In addition, animals fed the GM rice showed an altered immune response when challenged with an immunizing agent (sheep red blood cells).[45]

The authors went to great lengths to dismiss the large number of statistically significant differences, not only in GM rice composition, but also in the physiology and biochemistry of the animals on the GM rice diet. They claimed that these differences were within the range of natural variation or biologically not meaningful. However, an equally valid interpretation of the data is that the GM rice diet resulted in signs of harm to multiple systems of the body, which may escalate into serious harm if the study were extended beyond 90 days to the lifespan of the animals (2–3 years).

Altered gut bacteria and organ weights

Rats fed GM Bt rice for 90 days developed significant differences as compared with rats fed the non-GM isogenic line of rice. The GM-fed group had 23%

higher levels of coliform bacteria in their gut and there were differences in organ weights between the two groups, namely in the adrenals, testes, and uterus.[46]

Masking statistical significance through the concept of "biological relevance"

Study findings such as those described above have made it increasingly difficult for GM proponents to claim that there are no differences between the effects of GM foods and their non-GM counterparts. Clearly, there are.

To sidestep this problem, GM proponents have shifted their argument to claim that statistically significant effects are not "biologically relevant".

The concept of lack of biological relevance has been heavily promoted by the industry-funded group, the International Life Sciences Institute (ILSI), and affiliates to argue against regulatory restrictions on toxic chemicals.[49] But increasingly, it is invoked by authors defending the safety of GM crops[50] to argue that statistically significant observable effects in GM-fed animals are not important.

However, this argument is scientifically indefensible. Biological relevance with respect to changes brought about by GM foods has never been properly defined.

Most feeding trials on GM foods, including those carried out by industry to support applications for GM crop commercialization, are short- or medium-term studies of 30–90 days. These studies are too short to determine whether changes in animals fed a GM diet are biologically relevant or not.

In order to determine whether changes seen in these short- to medium-term studies are biologically relevant, the researchers would have to:

- Define in advance what "biological relevance" means with respect to effects found from feeding GM crops
- Extend the study duration from short- or medium-term to a long-term period. In the case of rodent studies, this would be two years – the major part of their lifespan[36]
- Examine the animals closely to see how any changes found in short- or medium-term studies progress – for example, they may disappear or lead to disease or premature death
- Analyze the biological relevance of the changes in light of the researchers' definition of the term

→ Carry out additional reproductive and multigenerational studies to determine effects on fertility and future generations.

Since these steps are not followed in cases where statistically significant effects are dismissed as not "biologically relevant", assurances of GM food safety founded on this line of argument are baseless.

In parallel with asserting lack of "biological relevance", a trend has grown of claiming that statistically significant effects of GM feed on experimental animals are not "adverse".[45] Again, however, the term "adverse" is not defined and the experiments are not extended to check whether any changes seen develop into more serious disease. So the term is technically meaningless.

EFSA responds to criticism over use of "biological relevance"

After being subjected to years of criticism by independent scientists and a member of the European Parliament over its use of "biological relevance",[36,51,52] in 2011 the European Food Safety Authority (EFSA) finally issued an Opinion on the relationship between statistical significance and biological relevance.[53]

But EFSA's Opinion fails to give a rigorous scientific or legal definition of what makes a statistically significant finding "biologically relevant" or not. Instead, it allows industry to come to its own conclusion on whether changes found in an experiment are "important", "meaningful", or "may have consequences for human health". These are vague concepts for which no measurable or objectively verifiable endpoints are defined. Thus they are a matter of opinion, not science.

The conclusions of the EFSA Opinion are not surprising, given that it is authored[53] by current or former affiliates of the industry-funded group, the International Life Sciences Institute (ILSI), including Harry Kuiper[54] (also then the chair of EFSA's GMO panel), Josef Schlatter, and Susan Barlow.[55,56] ILSI is funded by GM crop developer/agrochemical companies, including Monsanto.[57] Allowing ILSI affiliates to write EFSA's scientific advice on how to assess the safety of GM foods and crops is akin to allowing a student to write his or her own examination paper.

The "EU research shows GM foods are safe" argument

An EU research project is cited all over the world as evidence for GM crop and food safety. This project is of strategic importance to GMO promoters because the EU is often considered to be a GMO-skeptical region. The implication is that if the EU has decided GMOs are safe, then they must be.

However, the report based on this EU project, "A decade of EU-funded GMO research",[58] presents no such evidence.

Indeed, the project was not even designed to test the safety of any single GM food, but focuses on "the development of safety assessment approaches".[58] Taxpayers would be entitled to ask why the Commission spent €200 million of public money[58] on a research project that failed to address this most pressing of questions about GM foods.

In the SAFOTEST section of the report, which is dedicated to GM food safety, only five published animal feeding studies are referenced.[59,16,60,45,46] Two of these studies were carried out with a GM rice expressing a protein known to be toxic to mammals, in order to ascertain that the methodology used was sensitive enough to detect toxicity of a comparable level. These were so-called "positive control" experiments.[59,16] The other three studies were carried out with supposedly edible GM rice varieties that have not yet been commercialized anywhere in the world.[60,45,46]

None of the studies tested a commercialized GM food; none tested the GM food for long-term effects beyond the medium-term period of 90 days; all found differences between the animals fed GM and non-GM isogenic rice, which in some cases were statistically significant; and none concluded on the safety of the GM food tested, let alone on the safety of GM foods in general. Therefore the EU research project provides no evidence that could support claims of safety for any single GM food or for GM crops in general.

The "trillions of GM meals" argument

The argument that Americans have eaten trillions of GM meals with no ill effects is often repeated, but is unscientific and dishonest. No epidemiological studies have been carried out in the human population to track consumption of GM foods and to assess whether they might be causing ill effects. What is more, such studies are not possible on the continent where most GM meals are consumed – North America – as GM foods are not labelled there.

Unless consumption caused a rapid, acute, and obvious reaction that could be immediately traced back to a GM food, a link could not be found even if it did exist. An increase in incidence of a common, slow-developing disease like cancer, allergies, or kidney or liver damage would be difficult or impossible to link to GM foods.

The research of Dr Nancy Swanson and colleagues, based on official US government health statistics, shows that chronic illnesses have increased steeply in the US since GM crops were introduced. The trend correlates with rising glyphosate-based herbicide use on glyphosate-tolerant crops.[61,62] While this research does not prove causation, it does show that something is going badly wrong with the health of Americans. A link with GM crops and/or their associated pesticides cannot be ruled out.

References

1. Butelli E, Titta L, Giorgio M, et al. Enrichment of tomato fruit with health-promoting anthocyanins by expression of select transcription factors. Nat Biotechnol. 2008;26:1301-1308. doi:10.1038/nbt.1506.
2. Martin C. How my purple tomato could save your life. Mail Online. http://bit.ly/10JsmlO. Published November 8, 2008.
3. Heinemann J, Agapito-Tenfen SZ, Carman J. A comparative evaluation of the regulation of GM crops or products containing dsRNA and suggested improvements to risk assessments. Environ Int. 2013;55:43-55.
4. Zhang L, Hou D, Chen X, et al. Exogenous plant MIR168a specifically targets mammalian LDLRAP1: Evidence of cross-kingdom regulation by microRNA. Cell Res. 2012;22(1):107-126. doi:10.1038/cr.2011.158.
5. Bøhn T, Cuhra M, Traavik T, Sanden M, Fagan J, Primicerio R. Compositional differences in soybeans on the market: glyphosate accumulates in Roundup Ready GM soybeans. Food Chem. 2014;153(2014):207-215. doi:10.1016/j.foodchem.2013.12.054.
6. Gab-Alla AA, El-Shamei ZS, Shatta AA, Moussa EA, Rayan AM. Morphological and biochemical changes in male rats fed on genetically modified corn (Ajeeb YG). J Am Sci. 2012;8(9):1117-1123.
7. El-Shamei ZS, Gab-Alla AA, Shatta AA, Moussa EA, Rayan AM. Histopathological changes in some organs of male rats fed on genetically modified corn (Ajeeb YG). J Am Sci. 2012;8(10):684-696.
8. Hines FA. Memorandum to Linda Kahl on the Flavr Savr Tomato (Pathology Review PR–152; FDA Number FMF–000526): Pathology Branch's Evaluation of Rats with Stomach Lesions from Three Four-Week Oral (gavage) Toxicity Studies (IRDC Study Nos. 677–002, 677–004, and 677–005) and an Expert Panel's Report. US Department of Health & Human Services; 1993. http://www.biointegrity.org.
9. Pusztai A. Witness Brief – Flavr Savr tomato study in Final Report (IIT Research Institute, Chicago, IL 60616 USA) cited by Dr Arpad Pusztai before the New Zealand Royal Commission on Genetic Modification. 2000. http://www.gmcommission.govt.nz/.
10. Pusztai A. Can science give us the tools for recognizing possible health risks of GM food? Nutr Health. 2002;16:73-84.
11. Prescott VE, Campbell PM, Moore A, et al. Transgenic expression of bean alpha-amylase inhibitor in peas results in altered structure and immunogenicity. J Agric Food Chem. 2005;53:9023-9030. doi:10.1021/jf050594v.
12. Lee RY, Reiner D, Dekan G, Moore AE, Higgins TJV, Epstein MM. Genetically modified α-amylase inhibitor peas are not specifically allergenic in mice. PloS One. 2013;8:e52972. doi:10.1371/journal.pone.0052972.
13. Higgins TJ. Presentation at the GMSAFOOD Conference, Vienna, Austria, 6-8 March 2012 [video]; 2012. http://www.youtube.com/watch?v=yKda722clq8.
14. Higgins TJ. GMSAFOOD Project Press Conference [video]; 2012. http://www.youtube.com/watch?v=gHUKE_luMR8.
15. Finamore A, Roselli M, Britti S, et al. Intestinal and peripheral immune response to MON810 maize ingestion in weaning and old mice. J Agric Food Chem. 2008;56:11533-11539. doi:10.1021/jf802059w.
16. Kroghsbo S, Madsen C, Poulsen M, et al. Immunotoxicological studies of genetically modified rice expressing PHA-E lectin or Bt toxin in Wistar rats. Toxicology. 2008;245:24-34. doi:10.1016/j.tox.2007.12.005.
17. Krzyzowska M, Wincenciak M, Winnicka A, et al. The effect of multigenerational diet containing genetically modified triticale on immune system in mice. Pol J Vet Sci. 2010;13:423-430.
18. Tudisco R, Calabrò S, Cutrignelli MI, et al. Genetically modified soybean in a goat diet: Influence on kid performance. Small Rumin Res. 2015;126(Supplement 1):67-74. doi:10.1016/j.smallrumres.2015.01.023.
19. Malatesta M, Biggiogera M, Manuali E, Rocchi MBL, Baldelli B, Gazzanelli G. Fine structural analyses of pancreatic acinar cell nuclei from mice fed on genetically modified soybean. Eur J

Histochem. 2003;47:385-388.
20. Malatesta M, Caporaloni C, Gavaudan S, et al. Ultrastructural morphometrical and immunocytochemical analyses of hepatocyte nuclei from mice fed on genetically modified soybean. Cell Struct Funct. 2002;27:173-180.
21. Vecchio L, Cisterna B, Malatesta M, Martin TE, Biggiogera M. Ultrastructural analysis of testes from mice fed on genetically modified soybean. Eur J Histochem. 2004;48:448-454.
22. Malatesta M, Boraldi F, Annovi G, et al. A long-term study on female mice fed on a genetically modified soybean: effects on liver ageing. Histochem Cell Biol. 2008;130:967-977.
23. Séralini G-E, Clair E, Mesnage R, et al. Republished study: long-term toxicity of a Roundup herbicide and a Roundup-tolerant genetically modified maize. Environ Sci Eur. 2014;26(14). doi:10.1186/s12302-014-0014-5.
24. Hammond B, Dudek R, Lemen J, Nemeth M. Results of a 13 week safety assurance study with rats fed grain from glyphosate tolerant corn. Food Chem Toxicol. 2004;42:1003-1014. doi:10.1016/j.fct.2004.02.013.
25. Matthews J. Smelling a corporate rat. Spinwatch. http://www.spinwatch.org/index.php/issues/science/item/164-smelling-a-corporate-rat. Published December 11, 2012.
26. Hayes AW. Letter to Professor GE Séralini. November 2013. http://www.gmwatch.org/files/Letter_AWHayes_GES.pdf.
27. Hayes AW. Editor in chief of Food and Chemical Toxicology answers questions on retraction. Food Chem Toxicol. 2014. doi:10.1016/j.fct.2014.01.006.
28. EndScienceCensorship.org. Statement: Journal retraction of Séralini GMO study is invalid and an attack on scientific integrity. 2014. http://www.endsciencecensorship.org/en/page/Statement#.UwUSP14vFY4.
29. Institute of Science in Society. Open letter on retraction and pledge to boycott Elsevier. December 2013. http://www.i-sis.org.uk/Open_letter_to_FCT_and_Elsevier.php#form.
30. Schubert D. Science study controversy impacts world health. U-T San Diego. http://www.utsandiego.com/news/2014/jan/08/science-food-health/. Published January 8, 2014.
31. Robinson C. Was Séralini's republished paper peer-reviewed? GMWatch. http://gmwatch.org/index.php/news/archive/2014/15511. Published June 28, 2014.
32. Meyer H, Hilbeck A. Rat feeding studies with genetically modified maize - a comparative evaluation of applied methods and risk assessment standards. Environ Sci Eur. 2013;25(33). http://www.enveurope.com/content/pdf/2190-4715-25-33.pdf.
33. Brasil FB, Soares LL, Faria TS, Boaventura GT, Sampaio FJ, Ramos CF. The impact of dietary organic and transgenic soy on the reproductive system of female adult rat. Anat Rec Hoboken. 2009;292:587-594. doi:10.1002/ar.20878.
34. Carman JA, Vlieger HR, Ver Steeg LJ, et al. A long-term toxicology study on pigs fed a combined genetically modified (GM) soy and GM maize diet. J Org Syst. 2013;8:38-54.
35. Lynas M. GMO pigs study – more junk science. marklynas.org. June 2013. http://www.marklynas.org/2013/06/gmo-pigs-study-more-junk-science/.
36. Séralini GE, Mesnage R, Clair E, Gress S, de Vendômois JS, Cellier D. Genetically modified crops safety assessments: Present limits and possible improvements. Environ Sci Eur. 2011;23. doi:10.1186/2190-4715-23-10.
37. De Vendomois JS, Roullier F, Cellier D, Séralini GE. A comparison of the effects of three GM corn varieties on mammalian health. Int J Biol Sci. 2009;5:706-726.
38. Kilic A, Akay MT. A three generation study with genetically modified Bt corn in rats: Biochemical and histopathological investigation. Food Chem Toxicol. 2008;46:1164-1170. doi:10.1016/j.fct.2007.11.016.
39. US Food and Drug Administration (FDA). Biotechnology Consultation Note to the File BNF No 00077. Office of Food Additive Safety, Center for Food Safety and Applied Nutrition; 2002. http://bit.ly/ZUmiAF.
40. Trabalza-Marinucci M, Brandi G, Rondini C, et al. A three-year longitudinal study on the effects of a diet containing genetically modified Bt176 maize on the health status and performance of sheep. Livest Sci. 2008;113:178-190. doi:10.1016/j.livsci.2007.03.009.
41. Ewen SW, Pusztai A. Effect of diets containing genetically modified potatoes expressing Galanthus nivalis lectin on rat small intestine. Lancet. 1999;354:1353-1354. doi:10.1016/S0140-6736(98)05860-7.
42. Pusztai A, Bardocz S. GMO in animal nutrition: Potential benefits and risks. In: Mosenthin R, Zentek J, Zebrowska T, eds. Biology of Nutrition in Growing Animals. Vol 4. Elsevier Limited; 2006:513-540. http://www.sciencedirect.com/science/article/pii/S1877182309701043.
43. Fares NH, El-Sayed AK. Fine structural changes in the ileum of mice fed on delta-endotoxin-treated potatoes and transgenic potatoes. Nat Toxins. 1998;6(6):219-233.
44. European Food Safety Authority (EFSA). Scientific Opinion of the Panel on Genetically Modified Organisms on applications (EFSA-GMO-RX-MON810) for the renewal of authorisation for the continued marketing of (1) existing food and food ingredients produced from genetically modified insect resistant maize MON810; (2) feed consisting of and/or

containing maize MON810, including the use of seed for cultivation; and of (3) food and feed additives, and feed materials produced from maize MON810, all under Regulation (EC) No 1829/2003 from Monsanto. EFSA J. 2009;2009(1149):1-85.

45. Poulsen M, Kroghsbo S, Schroder M, et al. A 90-day safety study in Wistar rats fed genetically modified rice expressing snowdrop lectin Galanthus nivalis (GNA). Food Chem Toxicol. 2007;45:350-363. doi:10.1016/j.fct.2006.09.002.
46. Schrøder M, Poulsen M, Wilcks A, et al. A 90-day safety study of genetically modified rice expressing Cry1Ab protein (Bacillus thuringiensis toxin) in Wistar rats. Food Chem Toxicol. 2007;45:339-349. doi:10.1016/j.fct.2006.09.001.
47. Smith GDW. Is GM food good for you? Letter to the editor of the Sunday Times [unpublished]. March 2004.
48. Dona A, Arvanitoyannis IS. Health risks of genetically modified foods. Crit Rev Food Sci Nutr. 2009;49:164-175. doi:10.1080/10408390701855993.
49. Tyl RW, Crofton K, Moretto A, Moser V, Sheets LP, Sobotka TJ. Identification and interpretation of developmental neurotoxicity effects: a report from the ILSI Research Foundation/Risk Science Institute expert working group on neurodevelopmental endpoints. Neurotoxicol Teratol. 2008;30:349-381. doi:10.1016/j.ntt.2007.07.008.
50. Snell C, Aude B, Bergé J, et al. Assessment of the health impact of GM plant diets in long-term and multigenerational animal feeding trials: A literature review. Food Chem Toxicol. 2012;50(3–4):1134-1148.
51. Hilbeck A, Meier M, Römbke J, Jänsch S, Teichmann H, Tappeser B. Environmental risk assessment of genetically modified plants - concepts and controversies. Environ Sci Eur. 2011;23. doi:10.1186/2190-4715-23-13.
52. Breyer H. EFSA Definition of "Biological Relevance" in Connection with GMO Tests: Written Question by Hiltrud Breyer (Verts/ALE) to the Commission.; 2008. http://bit.ly/M6UFyn.
53. European Food Safety Authority (EFSA). Scientific opinion: Statistical significance and biological relevance. EFSA J. 2011;9:2372.
54. International Life Sciences Institute (ILSI). Nutritional and safety assessments of foods and feeds nutritionally improved through biotechnology, prepared by a task force of the ILSI International Food Biotechnology Committee. Compr Rev Food Sci Food Saf. 2004;3:38-104.
55. International Life Sciences Institute (ILSI). Risk Assessment of Genotoxic Carcinogens Task Force. Brussels, Belgium; 2011. http://bit.ly/1i8n8qk.
56. Constable A, Barlow S. Application of the Margin of Exposure Approach to Compounds in Food Which Are Both Genotoxic and Carcinogenic: Summary Report of a Workshop Held in October 2008 Organised by the ILSI Europe Risk Assessment of Genotoxic Carcinogens in Food Task Force. Brussels, Belgium: ILSI Europe; 2009. http://www.ilsi.org.ar/administrador/panel/newsletter/news27/moe_ws_report.pdf.
57. Corporate Europe Observatory. The International Life Sciences Institute (ILSI), a Corporate Lobby Group; 2012. http://corporateeurope.org/sites/default/files/ilsi-article-final.pdf.
58. European Commission Directorate-General for Research and Innovation, Biotechnologies, Agriculture, Food. A Decade of EU-Funded GMO Research (2001–2010). Brussels, Belgium; 2010. http://ec.europa.eu/research/biosociety/pdf/a_decade_of_eu-funded_gmo_research.pdf.
59. Poulsen M, Schrøder M, Wilcks A, et al. Safety testing of GM-rice expressing PHA-E lectin using a new animal test design. Food Chem Toxicol Int J Publ Br Ind Biol Res Assoc. 2007;45(3):364-377. doi:10.1016/j.fct.2006.09.003.
60. Knudsen I, Poulsen M. Comparative safety testing of genetically modified foods in a 90-day rat feeding study design allowing the distinction between primary and secondary effects of the new genetic event. Regul Toxicol Pharmacol. 2007;49(1):53-62. doi:10.1016/j.yrtph.2007.07.003.
61. Swanson NL, Leu A, Abrahamson J, Wallett B. Genetically engineered crops, glyphosate and the deterioration of health in the United States of America. J Org Syst. 2014;9(2). http://www.organic-systems.org/journal/92/JOS_Volume-9_Number-2_Nov_2014-Swanson-et-al.pdf.
62. Seneff S, Swanson N, Li C. Aluminum and glyphosate can synergistically induce pineal gland pathology: Connection to gut dysbiosis and neurological disease. Agric Sci. 2015;6:42-70

5 Myth: Many long-term studies show GM is safe

Truth: Few long-term studies have been carried out, but some show toxic effects from GM food

Myth at a glance

Some GMO proponents and scientists say that many long-term animal feeding studies have concluded GM foods are safe. But this is false. Few long-term and in-depth studies have been carried out and several studies that have been conducted have found toxic effects.

Long-term animal feeding studies are important because they can detect adverse health effects of a GMO diet that can take time to show up, such as organ damage and cancer. Most studies on GM foods are short- or medium-term. No regulatory authority anywhere in the world requires animal feeding studies beyond 90 days in rats (equivalent to only 7–8 years in a human).

The health effects detected in long-term animal feeding studies cannot be predicted merely by analyzing the composition of the GM food.

A review by Snell and colleagues purporting to present long-term studies showing long-term safety is misleading. Many of the studies are not long-term at all and the authors use double standards to dismiss findings of harm while accepting findings of safety at face value.

Long-term animal feeding studies are important because they can detect adverse health effects of a GMO diet that take time to show up, such as organ damage and cancer. Most studies on GM foods are short- or medium-term. No regulatory authority anywhere in the world requires animal feeding studies beyond 90 days in rats (equivalent to only 7–8 years in a human).

The health effects detected in long-term animal feeding studies cannot be predicted merely by analyzing the composition of the GM food.

Some GMO proponents and scientists say that many long-term animal feeding studies have concluded GM foods are safe. In fact, few long-term and in-depth studies have been carried out, but of those that have been carried out, some show toxic effects from the GM food. In addition, many studies claimed to be long-term are not. An analysis follows of some genuinely long-term studies and a review purporting to examine long-term studies on GMOs.

The Séralini study

This study found that a Monsanto GM Roundup-tolerant maize (NK603) and tiny amounts of the Roundup herbicide it was engineered to be grown with caused severe liver and kidney damage and hormonal disruption in rats fed over a long-term period of two years. Unexpected additional findings, which need to be followed up in a study with a larger number of rats, were an increased incidence of large palpable tumours, particularly those of the mammary gland in female animals, and premature death in some treatment groups.[1]

The Malatesta and Sakamoto studies

The extreme shortage of long-term studies on GM foods was highlighted by ANSES, France's food safety agency, in its critique[2] of the Séralini study on a GM herbicide-tolerant maize.[1] ANSES had conducted a search for long-term animal feeding studies on herbicide-tolerant GMOs, which make up over 80% of all commercialized GMOs[3] – to compare with Séralini's study. It found only two.[2,4]

One, by Manuela Malatesta's group in Italy, found health problems, including more acute signs of ageing in the liver, in mice fed GM glyphosate-tolerant soybeans.[5] The other, by Sakamoto and colleagues, concluded that there were no adverse effects in rats fed GM glyphosate-tolerant soybeans. However, glyphosate in the GM soy was only found at the limit of detection,[6] which suggests that the GM soy was not sprayed with glyphosate or that it was sprayed only once, early in the growing season. This does not reflect the normal way in which GM Roundup Ready soy is grown. A separate study found high levels of glyphosate residue in this type of GM soy grown under normal conditions.[7,8]

As a result of its analysis, ANSES called for more long-term studies on GM crops.[2]

The Snell review of supposedly long-term studies

A review by Snell and colleagues purports to examine the health impacts of GM foods as revealed by 24 long-term and multi-generational studies. The Snell review concludes that the GM foods examined are safe.[9] However, this cannot be justified by the papers discussed in the review.

Some of the studies examined by Snell and colleagues are not even toxicological studies that look at health effects. Instead they are so-called animal production studies that look at aspects of interest to food producers, such as feed conversion (the amount of weight the animal puts on relative to the amount of food it eats)[10] or milk production in cows.[11] Such studies do not examine health effects in detail and are not useful for assessing health risks of a GMO diet.

Several studies are not long-term at all and so cannot detect health effects that take time to show up, such as cancer and severe organ damage. For example, some studies in chickens last only for 35[12] and 42 days,[10] even though a chicken's natural lifespan is around seven years.

Snell and colleagues also categorized a 25-month feeding study with GM Bt maize in dairy cows by Steinke and colleagues[11] as long-term. But although most dairy cows are sent to slaughter at four to five years old because their productivity decreases after that age, a cow's natural lifespan is 17–20 years. A 25-month study in dairy cows is equivalent to around eight years in human terms or 90 days in rats. So from a toxicological point of view, Steinke's study could at most be described as medium-term (subchronic).

Moreover, in Steinke's study, half of the animals in the treatment group and the control group fell ill or proved infertile, for reasons that were not investigated or explained. In a scientifically unjustifiable move, these cows were simply removed from the study and replaced with other cows. No analysis is presented to show whether the problems that the cows suffered had anything to do with the diets tested.[11]

It is never acceptable to replace animals mid-way through a feeding experiment. For this reason alone, this study[11] is irrelevant to assessing health effects from GM feed and Snell and colleagues should not have included it in their analysis. That they did include it suggests that they had no meaningful quality control system in place, which in turn raises questions about the other studies cited.

Many of the studies reviewed were on animals that have a very different digestive system and metabolism to humans and are not relevant to assessing human health effects. These include studies on broiler chickens, cows, sheep, and fish.

Many of the studies tested only small groups of animals, raising the question of whether the study designs had sufficient statistical power to detect harm. Séralini's long-term experiment that found toxic effects from GM maize used 10 animals per sex per group[1] and was widely attacked for using too few animals for statistical significance[13] (though this criticism could only be justifiably levelled, at most, to the tumour and mortality observations, which normally require larger numbers of rats to evaluate). But Snell and colleagues happily cite experiments using 10 animals per group or less as evidence that GM foods are safe. This is an example of unscientific double standards, in which tests on 10 animals per sex per group are apparently sufficient to prove safety, but not enough to prove risk.

Some of the studies reviewed did find toxic effects in the GM-fed animals, but these were dismissed by Snell and colleagues. For example, findings of damage to liver and kidneys and alterations in blood biochemistry in rats fed GM Bt maize over three generations[14] were dismissed, as were the findings of Manuela Malatesta's team, of abnormalities in the liver, pancreatic, and testicular cells of mice fed on GM soy,[5,15,16,17] on the basis that the researchers used a non-isogenic soy variety as the non-GM comparator. This was unavoidable, given the refusal of GM seed companies to release their patented seeds to independent researchers.[18] Yet in an example of unscientific double standards, Snell and colleagues dismissed Malatesta's findings of risk, while accepting findings of safety in studies with precisely the same methodological weakness.

An objective assessment of Malatesta's findings would conclude that while the results do not show that GM soy was more toxic than the non-GM isogenic variety (because the isogenic variety was not used), they do show that GM herbicide-tolerant soy was more toxic than the non-GM soybeans tested, either because of the herbicide used, or the effect of the genetic engineering process, or a combination of the two.

In an extraordinary move, Snell and colleagues offered as the main counter to Malatesta's experimental findings a poster presentation with no references to any data, presented at a Society of Toxicology conference[19,20] by employees of the industry consultancy firm Exponent.[21] Though at first glance the reference given by Snell and colleagues has the appearance of a peer-reviewed paper, it is not.

Presentations given at conferences are not usually subjected to the scrutiny given to peer-reviewed publications. They certainly do not carry sufficient weight to counter original research findings from laboratory animal feeding experiments with GM foods, such as Malatesta's. This is particularly true when they do not base their argument on hard data, as in the case of the opinion piece cited by Snell and colleagues.

References

1. Séralini G-E, Clair E, Mesnage R, et al. Republished study: long-term toxicity of a Roundup herbicide and a Roundup-tolerant genetically modified maize. Environ Sci Eur. 2014;26(14). doi:10.1186/s12302-014-0014-5.
2. ANSES (French Agency for Food Environmental and Occupational Health & Safety). Opinion Concerning an Analysis of the Study by Séralini et Al. (2012) "Long Term Toxicity of a ROUNDUP Herbicide and a ROUNDUP-Tolerant Genetically Modified Maize."; 2012. http://www.anses.fr/sites/default/files/files/BIOT2012sa0227EN.pdf.
3. James C. Global Status of Commercialized biotech/GM Crops: 2012. Ithaca, NY: ISAAA; 2012. http://www.isaaa.org/resources/publications/briefs/44/download/isaaa-brief-44-2012.pdf.
4. ANSES (French Agency for Food Environmental and Occupational Health & Safety). ANSES highlights the weaknesses of the study by Séralini et al., but recommends new research on the long-term effects of GMOs. 2012. http://www.anses.fr/en/content/anses-highlights-weaknesses-study-s%C3%A9ralini-et-al-recommends-new-research-long-term-effects.
5. Malatesta M, Boraldi F, Annovi G, et al. A long-term study on female mice fed on a genetically modified soybean: effects on liver ageing. Histochem Cell Biol. 2008;130:967-977.
6. Sakamoto Y, Tada Y, Fukumori N, et al. [A 104-week feeding study of genetically modified soybeans in F344 rats]. Shokuhin Eiseigaku Zasshi J Food Hyg Soc Jpn. 2008;49:272-282.
7. Bøhn T, Cuhra M, Traavik T, Sanden M, Fagan J, Primicerio R. Compositional differences in soybeans on the market: glyphosate accumulates in Roundup Ready GM soybeans. Food Chem. 2013;153(2014):207-215. doi:10.1016/j.foodchem.2013.12.054.
8. Bøhn T, Cuhra M. How "extreme levels" of Roundup in food became the industry norm. Indep Sci News. 2014. http://www.independentsciencenews.org/news/how-extreme-levels-of-roundup-in-food-became-the-industry-norm/.
9. Snell C, Aude B, Bergé J, et al. Assessment of the health impact of GM plant diets in long-term and multigenerational animal feeding trials: A literature review. Food Chem Toxicol. 2012;50(3–4):1134-1148.
10. Brake J, Faust MA, Stein J. Evaluation of transgenic event Bt11 hybrid corn in broiler chickens. Poult Sci. 2003;82:551-559.
11. Steinke K, Guertler P, Paul V, et al. Effects of long-term feeding of genetically modified corn (event MON810) on the performance of lactating dairy cows. J Anim Physiol Anim Nutr Berl. 2010;94:e185-e193. doi:10.1111/j.1439-0396.2010.01003.x.
12. Flachowsky G, Aulrich K, Böhme H, Halle I. Studies on feeds from genetically modified plants (GMP) – Contributions to nutritional and safety assessment. Anim Feed Sci Technol. 2007;133:2-30.
13. AFP. Controversial Seralini study linking GM to cancer in rats is republished. The Guardian (UK). http://www.theguardian.com/environment/2014/jun/24/controversial-seralini-study-gm-cancer-rats-republished. Published June 24, 2014.
14. Kilic A, Akay MT. A three generation study with genetically modified Bt corn in rats: Biochemical and histopathological investigation. Food Chem Toxicol. 2008;46:1164-1170. doi:10.1016/j.fct.2007.11.016.
15. Malatesta M, Biggiogera M, Manuali E, Rocchi MBL, Baldelli B, Gazzanelli G. Fine structural analyses of pancreatic acinar cell nuclei from mice fed on genetically modified soybean. Eur J Histochem. 2003;47:385-388.
16. Malatesta M, Caporaloni C, Gavaudan S, et al. Ultrastructural morphometrical and immunocytochemical analyses of hepatocyte nuclei from mice fed on genetically modified soybean. Cell Struct Funct. 2002;27:173-180.
17. Vecchio L, Cisterna B, Malatesta M, Martin TE, Biggiogera M. Ultrastructural analysis of testes from mice fed on genetically modified soybean. Eur J Histochem. 2004;48:448-454.
18. Waltz E. Under wraps – Are the crop industry's strong-arm tactics and close-fisted attitude to sharing seeds holding back independent research and undermining public acceptance of transgenic crops? Nat Biotechnol. 2009;27(10):880-882. doi:10.1038/nbt1009-880.
19. Williams AL, DeSesso JM. Genetically-modified soybeans: A critical evaluation of studies addressing potential changes associated with ingestion; Abstract 1154, Poster Board 424, Safety Concerns of Food and Natural Products. In: ; 2010. http://www.exponent.com/files/Uploads/Documents/news%20and%20features/SOT%20Presentation%20Handout_draft.pdf.
20. Williams AL, DeSesso JM. Genetically-modified soybeans: A critical evaluation of studies addressing potential changes associated with ingestion; Abstract 1154. The Toxicologist. 2010;114:246.
21. Williams AL, Watson RE, DeSesso JM. Developmental and reproductive outcomes in humans and animals after glyphosate exposure: A critical analysis. J Toxicol Environ Health Part B. 2012;15:39-96.

6 Myth: The Nicolia review compiles 1,700+ studies showing that GMOs are safe

Truth: Many of the papers fail to document GMO safety; some show that certain GMOs are harmful; and important papers relevant to GMO safety are omitted or ignored

Myth at a glance

A review by Nicolia and colleagues is often used to claim that 1700+ studies show GMOs are safe.

But this is false. Most of the studies listed by Nicolia and colleagues do not provide any hard data on the safety of GMOs for health or the environment.

Some of the studies listed actually show the GMO tested is harmful and others show risks from GMOs.

Many studies providing useful data on GMO safety are simply omitted from consideration by the review authors.

In addition, the review misleadingly reports a major scientific controversy around the effects of Bt toxins in GM Bt crops on non-target organisms. The scientifically defensible conclusion that can be drawn from the experimental literature on this topic is that Bt toxins have harmful effects on non-target organisms.

A review by Nicolia and colleagues, "An overview of the last 10 years of genetically engineered crop safety research",[1] is widely cited[2,3] to argue that over 1,700 studies show GM crops and foods are safe for human and animal health and the environment. However, only a relatively small proportion of the studies cited in the Nicolia review and the supplementary list of 1,700 papers actually address the questions of GMO safety.

The rest of the studies cited in the Nicolia review are irrelevant or tangential to an investigation of the safety of GM foods and crops. Irrelevant categories of studies include the following (examples of studies included in the review and supplementary materials are referenced):

- Animal production studies, often performed by GM companies on their own products.[4,5] These do not assess in detail the health impacts of GM foods but look at aspects of animal production of interest to the food and agriculture industry, such as weight gain and milk production. Many of these studies are on animals that have very different digestive systems and metabolisms from humans; and most are short-term in comparison to the animals' natural lifespan. They do not provide useful information on long-term or human health risks.
- Opinion pieces, advocacy articles, and reviews of safety assessment approaches and regulations relating to GMOs.[6,7,8,9,10] These papers do not provide any original data and contribute nothing of substance to the debate on safety.
- Studies on experimental GM crops that have never been commercialized.[11,12] Some of these studies reveal the imprecise and unpredictable nature of GM technology because they show unintended differences between the GM crop and the non-GM parent[12] or toxic effects in animals exposed to the GM crop.[11] However, because each GM transformation event is different, the findings of such studies are not useful in assessing the safety of the GMOs already in our food and feed supplies.
- Studies on consumer perceptions of GM foods.[13]

These studies do not provide evidence on the risks posed by GMOs to human or animal health or the environment. Moreover, Nicolia and colleagues omit from their list, or from their discussion, important studies that find risks and toxic effects from GMOs. This is in spite of the fact that these studies are seminal to any discussion of GMO safety.[14,15,16,17,18,19,20,21] One glaring omission in Nicolia and colleagues' review (though included in the supplementary list) is the detailed research of Manuela Malatesta, which found toxic effects, including more acute signs of ageing in the liver, in mice fed GM soy over a long-term period.[15,16,17,18,19]

Nicolia and colleagues mislead on "substantial equivalence"

Nicolia and colleagues incorrectly claim that there is a "consensus" about the validity of the concept of substantial equivalence in GMO risk assessment. The concept of substantial equivalence has been heavily criticized and challenged from the beginning by independent scientists because it has never been scientifically or legally defined.[22,23,24,25,26,27,28,29] In practice, there can be substantial compositional differences in the GMO compared with the non-GM comparator crop, but the GMO developer company still declares the GMO as "equivalent" and regulators accept the designation.

Nicolia and colleagues admit that the literature on substantial equivalence is mostly composed of papers produced by GMO companies, but fail to draw the obvious conclusion that only an appearance of consensus has been generated due to the dominance in the literature of this biased group of authors.

Nicolia and colleagues mislead on major controversy over Bt crop safety

Another example of Nicolia and colleagues' misleading reporting is their treatment of the literature on toxic effects of GM Bt crops on non-target species. They claim this literature "shows little or no evidence of the negative effects of GE crops."

But they only reach this conclusion by ignoring important papers and misrepresenting the evidence in others. Assisted by their 10-year cut-off date (in a scientifically unjustifiable move, they only include studies from the past ten years), they misleadingly report a major scientific controversy around the effects of Bt toxins on non-target organisms.

The controversy began in the mid-1990s, when studies led by Dr Angelika Hilbeck showed that Bt toxins of microbial and Bt plant origin caused lethal effects in the larvae of the green lacewing, a beneficial insect to farmers, when administered using a protocol that ensured ingestion.[30,31,32] A separate study also led by Hilbeck (Schmidt and colleagues, 2009) found that Bt toxins caused increased mortality in the larvae of another beneficial insect, the ladybird.[33]

Based on this study and over 30 others, in 2009 Germany banned the cultivation of Monsanto's Bt maize MON810,[34] which contains one of the Bt toxins that Hilbeck's team found to be harmful.[33]

Nicolia and colleagues included the Hilbeck ladybird study[33] in their supplementary list of 1,700 articles, but ignored its findings in their discussion in their main review paper.

Rebuttal studies were carried out, apparently to disprove the findings of Hilbeck's teams and undermine the German ban. These studies affirmed the safety of Bt toxins for lacewings[35,36,37] and ladybirds.[38,39,40] The authors of the rebuttal study on ladybirds (Alvarez-Alfageme and colleagues) found no ill effects on ladybird larvae fed Bt toxins and said that the "apparent harmful effects" found by Schmidt and colleagues were due to "poor study design and procedures".[38]

Nicolia and colleagues included several of the rebuttal studies in their list of 1,700. However, they failed to cite follow-up studies by Hilbeck and colleagues that proved that the rebuttal studies were poorly designed and executed. Hilbeck and colleagues proved that changes in the testing protocols were the reasons for the rebuttal studies failing to find the same results of toxic effects to lacewings and ladybirds.

Hilbeck and colleagues showed that the lacewings in the rebuttal study could not have ingested the Bt toxins in the form provided, as their mouthparts are formed in such a way as to make ingestion impossible.[34] This is equivalent to testing an oral drug for side-effects by providing it in pills too large for the subjects to swallow.

Hilbeck and colleagues did further experiments[34,41] to test the claims of the rebuttal study on ladybird larvae (Alvarez-Alfageme et al, 2011[38]). Again, the results of the rebuttal study were shown to be due to altered and inadequate protocols. Hilbeck and colleagues repeated Alvarez-Alfageme's methodology – and found that the water in the sugar solution in which the Bt toxin had been fed completely evaporated after a few hours, making it unlikely that the larvae in Alvarez-Alfageme's experiment even ingested the Bt toxin. When Hilbeck and colleagues made the Bt toxins available continuously in a way that the ladybird larvae could access, a lethal effect on the larvae was found.[41]

Unaccountably, Nicolia and colleagues omit the two confirmatory empirical studies by Hilbeck's team[34,41] even from their list of 1,700 studies, although they fall within their 10-year timeframe. They entirely ignore the scientific demolition by Hilbeck's team of the flawed rebuttal studies.

This is just one example among many showing that Nicolia and colleagues' review is misleading.

References

1. Nicolia A, Manzo A, Veronesi F, Rosellini D. An overview of the last 10 years of genetically engineered crop safety research. Crit Rev Biotechnol. 2013:1-12. doi:10.3109/07388551.2013.823595.
2. Bailey P. GMOs are nothing to fear. US News & World Report. http://www.usnews.com/opinion/articles/2013/11/04/scientific-evidence-doesnt-show-gmos-are-harmful. Published November 4, 2013.
3. Wendel J. With 2000+ global studies confirming safety, GM foods among most analyzed subjects in science. Genetic Literacy Project. http://bit.ly/1bjhPQG. Published October 8, 2013.

4. Taylor M, Hartnell G, Lucas D, Davis S, Nemeth M. Comparison of broiler performance and carcass parameters when fed diets containing soybean meal produced from glyphosate-tolerant (MON 89788), control, or conventional reference soybeans. Poult Sci. 2007;86(12):2608-2614. doi:10.3382/ps.2007-00139.
5. Bakke-McKellep AM, Sanden M, Danieli A, et al. Atlantic salmon (Salmo salar L.) parr fed genetically modified soybeans and maize: Histological, digestive, metabolic, and immunological investigations. Res Vet Sci. 2008;84:395-408.
6. Kok EJ, Keijer J, Kleter GA, Kuiper HA. Comparative safety assessment of plant-derived foods. Regul Toxicol Pharmacol. 2008;50:98-113. doi:10.1016/j.yrtph.2007.09.007.
7. Kok EJ, Kuiper HA. Comparative safety assessment for biotech crops. Trends Biotechnol. 2003;21:439-444.
8. Chassy BM. Food safety evaluation of crops produced through biotechnology. J Am Coll Nutr. 2002;21(3 Suppl):166S - 173S.
9. Konig A, Cockburn A, Crevel RW, et al. Assessment of the safety of foods derived from genetically modified (GM) crops. Food Chem Toxicol. 2004;42:1047-1088. doi:10.1016/j.fct.2004.02.019.
10. International Life Sciences Institute (ILSI). Nutritional and safety assessments of foods and feeds nutritionally improved through biotechnology, prepared by a task force of the ILSI International Food Biotechnology Committee. Compr Rev Food Sci Food Saf. 2004;3:38-104.
11. Kroghsbo S, Madsen C, Poulsen M, et al. Immunotoxicological studies of genetically modified rice expressing PHA-E lectin or Bt toxin in Wistar rats. Toxicology. 2008;245:24-34. doi:10.1016/j.tox.2007.12.005.
12. Poulsen M, Kroghsbo S, Schroder M, et al. A 90-day safety study in Wistar rats fed genetically modified rice expressing snowdrop lectin Galanthus nivalis (GNA). Food Chem Toxicol. 2007;45:350-363. doi:10.1016/j.fct.2006.09.002.
13. Magnusson MK, Koivisto Hursti U-K. Consumer attitudes towards genetically modified foods. Appetite. 2002;39(1):9-24. doi:10.1006/appe.2002.0486.
14. Domingo JL, Bordonaba JG. A literature review on the safety assessment of genetically modified plants. Env Int. 2011;37:734-742.
15. Malatesta M, Caporaloni C, Gavaudan S, et al. Ultrastructural morphometrical and immunocytochemical analyses of hepatocyte nuclei from mice fed on genetically modified soybean. Cell Struct Funct. 2002;27:173-180.
16. Malatesta M, Caporaloni C, Rossi L, et al. Ultrastructural analysis of pancreatic acinar cells from mice fed on genetically modified soybean. J Anat. 2002;201:409-415.
17. Malatesta M, Biggiogera M, Manuali E, Rocchi MBL, Baldelli B, Gazzanelli G. Fine structural analyses of pancreatic acinar cell nuclei from mice fed on genetically modified soybean. Eur J Histochem. 2003;47:385-388.
18. Malatesta M, Boraldi F, Annovi G, et al. A long-term study on female mice fed on a genetically modified soybean: effects on liver ageing. Histochem Cell Biol. 2008;130:967-977.
19. Vecchio L, Cisterna B, Malatesta M, Martin TE, Biggiogera M. Ultrastructural analysis of testes from mice fed on genetically modified soybean. Eur J Histochem. 2004;48:448-454.
20. Kilic A, Akay MT. A three generation study with genetically modified Bt corn in rats: Biochemical and histopathological investigation. Food Chem Toxicol. 2008;46:1164-1170. doi:10.1016/j.fct.2007.11.016.
21. Dona A, Arvanitoyannis IS. Health risks of genetically modified foods. Crit Rev Food Sci Nutr. 2009;49:164-175. doi:10.1080/10408390701855993.
22. Séralini G-E, Clair E, Mesnage R, et al. Republished study: long-term toxicity of a Roundup herbicide and a Roundup-tolerant genetically modified maize. Environ Sci Eur. 2014;26(14). doi:10.1186/s12302-014-0014-5.
23. De Vendomois JS, Roullier F, Cellier D, Séralini GE. A comparison of the effects of three GM corn varieties on mammalian health. Int J Biol Sci. 2009;5:706-726.
24. Then C, Bauer-Panskus A. European Food Safety Authority: A Playing Field for the Biotech Industry. Munich, Germany: Testbiotech; 2010. http://www.testbiotech.de/en/node/431.
25. Levidow L, Murphy J, Carr S. Recasting "substantial equivalence": Transatlantic governance of GM food. Sci Technol Hum Values. 2007;32:26-64.
26. Millstone E, Brunner E, Mayer S. Beyond "substantial equivalence." Nature. 1999;401:525-526. doi:10.1038/44006.
27. Pusztai A. Genetically modified foods: Are they a risk to human/animal health? Actionbioscience.org. http://www.actionbioscience.org/biotech/pusztai.html. Published June 2001.
28. Pusztai A, Bardocz S. GMO in animal nutrition: Potential benefits and risks. In: Mosenthin R, Zentek J, Zebrowska T, eds. Biology of Nutrition in Growing Animals.Vol 4. Elsevier Limited; 2006:513-540. http://www.sciencedirect.com/science/article/pii/S1877182309701043.
29. Lotter D. The genetic engineering of food and the failure of science – Part 1: The development of a flawed enterprise. Int Jrnl Soc Agr Food. 2008;16:31-49.

30. Hilbeck A, Moar WJ, Pusztai-Carey M, Filippini A, Bigler F. Toxicity of Bacillus thuringiensis Cry1Ab toxin to the predator Chrysoperla carnea (Neuroptera: Chrysopidae). Environ Entomol. 1998;27(5):1255-1263.
31. Hilbeck A, Moar WJ, Pusztai-Carey M, Filippini A, Bigler F. Prey-mediated effects of Cry1Ab toxin and protoxin and Cry2A protoxin on the predator Chrysoperla carnea. Entomol Exp Appl. 1999;91:305-316.
32. Hilbeck A, Baumgartner M, Fried PM, Bigler F. Effects of transgenic Bt corn-fed prey on immature development of Chrysoperla carnea (Neuroptera: Chrysopidae). Environ Entomol. 1998;27(2):480-487.
33. Schmidt JE, Braun CU, Whitehouse LP, Hilbeck A. Effects of activated Bt transgene products (Cry1Ab, Cry3Bb) on immature stages of the ladybird Adalia bipunctata in laboratory ecotoxicity testing. Arch Env Contam Toxicol. 2009;56(2):221-228. doi:10.1007/s00244-008-9191-9.
34. Hilbeck A, Meier M, Trtikova M. Underlying reasons of the controversy over adverse effects of Bt toxins on lady beetle and lacewing larvae. Environ Sci Eur. 2012;24(9). doi:10.1186/2190-4715-24-9.
35. Dutton A, Klein H, Romeis J, Bigler F. Uptake of Bt-toxin by herbivores feeding on transgenic maize and consequences for the predator Chrysoperla carnea. Ecol Entomol. 2002;27:441-447.
36. Dutton A, Klein H, Romeis J, Bigler F. Prey-mediated effects of Bacillus thuringiensis spray on the predator Chrysoperla carnea in maize. Biol Control. 2003;26(2):209-215. doi:10.1016/S1049-9644(02)00127-5.
37. Romeis J, Dutton A, Bigler F. Bacillus thuringiensis toxin (Cry1Ab) has no direct effect on larvae of the green lacewing Chrysoperla carnea (Stephens) (Neuroptera: Chrysopidae). J Insect Physiol. 2004;50(2-3):175-183. doi:10.1016/j.jinsphys.2003.11.004.
38. Alvarez-Alfageme F, Bigler F, Romeis J. Laboratory toxicity studies demonstrating no adverse effects of Cry1Ab and Cry3Bb1 to larvae of Adalia bipunctata (Coleoptera: Coccinellidae): the importance of study design. Transgenic Res. 2011;20:467-479.
39. Rauschen S. A case of "pseudo science"? A study claiming effects of the Cry1Ab protein on larvae of the two-spotted ladybird is reminiscent of the case of the green lacewing. Transgenic Res. 2010;19:13-16. doi:10.1007/s11248-009-9301-0.
40. Ricroch A, Berge JB, Kuntz M. Is the German suspension of MON810 maize cultivation scientifically justified? Transgenic Res. 2010;19:1-12. doi:10.1007/s11248-009-9297-5.
41. Hilbeck A, McMillan JM, Meier M, Humbel A, Schlaepfer-Miller J, Trtikova M. A controversy re-visited: Is the coccinellid Adalia bipunctata adversely affected by Bt toxins? Environ Sci Eur. 2012;24(10). doi:10.1186/2190-4715-24-10.

7 Myth: The Van Eenennaam review considered data from 100 billion animals and found GMOs are safe

Truth: The review provides no data proving GMO safety

Myth at a glance

The Van Eenennaam and Young review (2014) has been cited as showing GMO safety. The review considered 29 years of livestock productivity field data in the US from before and after the introduction of GMOs, representing more than 100 billion farm animals. These data do not show that livestock health declined after the introduction of GM feed and thus are said to show that GMOs cause no health problems in animals.

However, the review tells us very little about GMO safety. The data from the "100 billion animals" are uncontrolled for many variables, including escalating antibiotic use in livestock, which can mask health problems. It is not even known how many of these animals were eating GMOs and for how long. An astonishing 98.45% of the 100 billion farm animals cited by Van Eenennaam are poultry, with 92.17% being broiler chickens. These animals are irrelevant models for assessing human or even mammalian health risks.

Also, broiler chickens live for a maximum of 49 days before being sent to slaughter, a small fraction of a chicken's natural lifespan. So these data give no information on long-term health effects of GMOs in poultry and no information on any types of health effects, short- or long-term, in mammals, including humans.

The Van Eenennaam and Young review (2014), authored by a former Monsanto scientist,[1] has been cited as a "trillion-meal study" that shows GMO safety.[2] The review considered 29 years of livestock productivity field data in the US from before and after the introduction of GMOs. These data are said to represent more than 100 billion animals and to show that livestock health did not decline after the introduction of GM feed. The conclusion is drawn that GMOs cause no health problems in animals.

However, these data are irrelevant to assessing the health risks of GMO consumption in humans or even in animals. The safety of GM food for animal consumption can only be assessed through controlled animal feeding studies in which one group of animals is fed a GM diet and the other group of animals a non-GM diet. "Controlled" means that the study tests one variable at a time: all aspects of the animals' treatment and diet must be the same except for the genetic modification. For the most part, Van Eenennaam's review does not provide data of this type.

The vast majority of the data that are claimed to represent the 100 billion farm animals were not generated in a controlled study but are taken from livestock "field data". Many things have changed in livestock husbandry in recent decades, including improved genetics (artificial insemination with specially selected stud males) and escalating antibiotic use, which can hide inflammation and other health problems. The review does not control for these and other factors.

Unusually for a study claiming to address health effects, it is not until pages 37–38 of the study that the authors present their data justifying the 100 billion animals claim. Problems with these data include:

- No details of what the animals ate (percentage of GM feed to non-GM) are given. This is especially misleading in the case of cattle, which may be raised predominantly on (non-GMO) grass and silage before being "finished" on grain, an unspecified proportion of which may be GMO, in feedlots.
- An astonishing 98.45% of the 100 billion farm animals cited by Van Eenennaam are poultry, with 92.17% being broiler chickens. These animals are irrelevant models for assessing human health risks since they have different digestive systems and metabolism.
- By the authors' own admission, broiler chickens live for a maximum of 49 days before being sent to slaughter, a small fraction of a chicken's natural lifespan of 7 years. Thus these data give no information on long-term health effects of GMOs in poultry and no information on any types of health effects, short- or long-term, in humans.

- An additional small fraction of Van Eenennaam's data set (3.63%) consists of cattle. These cattle, as stated above, have a diet with an unknown proportion of GM feed, which is fed for an unknown duration. Cattle also have short commercial lifespans in relation to their natural lifespan: 4–5 years for dairy cattle and 24 months for beef cattle, versus a 17–20-year natural lifespan. Again, these data say nothing about long-term effects.
- Just 1.19% of Van Eenennaam's dataset is drawn from an animal model that is considered relevant to human health: pigs. However, the same problem of short-term data applies: pigs have a short commercial lifespan of 5–6 months versus a natural lifespan of 10–12 years. So the dataset contains no long-term data.
- Many health problems in livestock do not show up in slaughterhouse inspections and thus will not be included in Van Eenennaam's dataset. For example, Carman and colleagues (2013)[3] found a higher rate of severe stomach inflammation and heavier uteri in pigs fed a GMO diet compared with pigs fed a non-GMO diet, but all the pigs in this study passed the slaughterhouse inspection.[4] Because they were not "condemned" at slaughter, they would have been included in Van Eenennaam's dataset as normal, healthy animals.
- Other serious problems associated by farmers with GM feed may not show up in Van Eenennaam's dataset. For example, the Danish pig farmer Ib Pedersen reported reduced litter size, fewer live-born piglets, increased need for medication, increased diarrhoea and bloat, and malformations in pigs fed GM glyphosate-tolerant soy – problems that disappeared or decreased when he switched to non-GMO feed.[5] Van Eenennaam does not give enough information to enable us to know which, if any, of these types of health problems are included in her dataset, and medication use is not considered.

These points serve to emphasize the importance of using peer-reviewed data from controlled studies rather than the uncontrolled "field data" used by Van Eenennaam and Young. Using uncontrolled data to make claims that GMO foods are safe to eat is equivalent to considering a population in which the birth rate is rising and the incidence of smoking is also rising, and concluding that smoking is safe for mothers and foetuses. Scientifically, it is nonsense.

Review cites problematic peer-reviewed studies

Van Eenennaam and Young cite some peer-reviewed animal feeding studies to back their claims of safety. They single out two for special praise on the grounds that they are long-term, "thorough", and used "appropriate controls". However, closer examination of these studies shows that they offer no reassurance about GMO safety. The first is a study on dairy cows by Steinke et al (2010) with MON810 maize.[6,7,8] But as explained in Chapter 5, this study is not long-term and the researchers replaced half the animals at an unspecified point in the study, making the results worthless.

The second study was on pigs fed GM MON810 Bt maize.[9,10,11,12,13] Van Eenennaam and Young say no "long-term adverse effects" were found in this study. However, this is to quote the findings selectively. Bt maize was found to trigger an immune response when fed over a short period of 31 days.[12] The researchers expanded the study to a longer period and assessed the immune functioning of sows fed Bt maize for 143 days through gestation and lactation, and of their offspring.

It is important to note that even this study period is not long-term, since pigs can live for 10–12 years. Nonetheless immune disturbances and changes in blood biochemistry were found in the GM-fed group, but the authors dismissed them, concluding, "Treatment differences observed following feeding of Bt maize to sows did not indicate inflammation or allergy and are unlikely to be of major importance."[11]

While the data derived from this experiment can be seen as objective, the conclusion that the changes seen "are unlikely to be of major importance" is a matter of subjective interpretation. Other scientists looking at the same data may conclude that immune disturbances and blood changes are of concern because their significance is not fully understood and because they may lead in the longer term to allergic-type reactions or compromised immune system function.

What is not in doubt is that the pig study shows that GM MON810 maize is not equivalent to the non-GM isogenic parent maize in its effects and that metabolically and biologically, MON810 maize has different effects than the non-GM isogenic parent. Scientists can debate whether these changes are adverse or biologically significant, but until those questions are resolved with further studies, it is not valid to conclude from this study that either the GMO tested is safe, or that GMOs in general are safe.

References

1. US Food and Drug Administration (FDA). Alison Louise Van Eenennaam. Undated. http://www.fda.gov/downloads/AdvisoryCommittees/CommitteesMeetingMaterials/VeterinaryMedicineAdvisoryCommittee/UCM225072.pdf.
2. Entine J. The debate about GMO safety Is over, thanks to a new trillion-meal study. Forbes. 2014. http://www.forbes.com/sites/jonentine/2014/09/17/the-debate-about-gmo-safety-is-over-thanks-to-a-new-trillion-meal-study/.
3. Carman JA, Vlieger HR, Ver Steeg LJ, et al. A long-term toxicology study on pigs fed a combined genetically modified (GM) soy and GM maize diet. J Org Syst. 2013;8:38-54.
4. Carman J. Personal communication with the authors. 2014.
5. Pedersen IB. Changing from GMO to non-GMO natural soy, experiences from Denmark. Science in Society. September 10, 2014.
6. Steinke K, Guertler P, Paul V, et al. Effects of long-term feeding of genetically modified corn (event MON810) on the performance of lactating dairy cows. J Anim Physiol Anim Nutr Berl. 2010;94:e185-e193. doi:10.1111/j.1439-0396.2010.01003.x.
7. Guertler P, Paul V, Steinke K, et al. Long-term feeding of genetically modified corn (MON810) — Fate of cry1Ab DNA and recombinant protein during the metabolism of the dairy cow. Livest Sci. 2010;131(2–3):250-259. doi:10.1016/j.livsci.2010.04.010.
8. Guertler P, Brandl C, Meyer HHD, Tichopad A. Feeding genetically modified maize (MON810) to dairy cows: comparison of gene expression pattern of markers for apoptosis, inflammation and cell cycle. J Verbr Leb. 2012;7:195-202.
9. Buzoianu SG, Walsh MC, Rea MC, et al. High-throughput sequence-based analysis of the intestinal microbiota of weanling pigs fed genetically modified MON810 maize expressing Bacillus thuringiensis Cry1Ab (Bt maize) for 31 days. Appl Env Microbiol. 2012;78:4217-4224. doi:10.1128/AEM.00307-12.
10. Buzoianu SG, Walsh MC, Rea MC, et al. The effect of feeding Bt MON810 maize to pigs for 110 days on intestinal microbiota. PLoS ONE. 2012;7:e33668. doi:10.1371/journal.pone.0033668.
11. Buzoianu SG, Walsh MC, Rea MC, et al. Effects of feeding Bt maize to sows during gestation and lactation on maternal and offspring immunity and fate of transgenic material. PLoS ONE. 2012;7(10). doi:10.1371/journal.pone.0047851.
12. Walsh MC, Buzoianu SG, Gardiner GE, et al. Fate of transgenic DNA from orally administered Bt MON810 maize and effects on immune response and growth in pigs. PLoS ONE. 2011;6:e27177. doi:10.1371/journal.pone.0027177.
13. Buzoianu SG, Walsh MC, Rea MC, et al. Effect of feeding genetically modified Bt MON810 maize to ~40-day-old pigs for 110 days on growth and health indicators. Anim Int J Anim Biosci. 2012;6(10):1609-1619. doi:10.1017/S1751731112000249.

8 Myth: GM crops increase yield potential

Truth: GM crops do not increase yield potential

Myth at a glance

GM has not increased the yield potential of crops. Though yield increases were seen in major crops in the twentieth century, these were due to improvements from conventional breeding, not to GM traits. In some cases GM crops yield less than non-GM crops.

High yield is a complex genetic trait based on multiple gene functions and cannot be genetically engineered into a crop. Virtually all commercially available GM traits confer tolerance to herbicides or insect resistance, or both. High-yielding GM crop varieties owe their high yields to conventional breeding. Herbicide-tolerant or insect-resistant GM traits are then added to the high-yielding non-GM varieties.

Claims that GM Bt cotton has improved cotton yields in India do not stand up to scrutiny, since the yield increase occurred before Bt cotton was widely introduced. Since Bt cotton has come to dominate Indian production, yields have declined.

GM crops are often claimed to give higher yields than naturally bred varieties. But the data do not support this claim. GM crops have not performed consistently better than their non-GM counterparts, with GM soybeans giving lower yields in university-based trials.[1,2]

Controlled field trials comparing GM and non-GM soy production suggested that 50% of the drop in yield was due to the disruption in gene function caused by the GM transformation process.[2] Similarly, field tests of Bt maize hybrids showed that they took longer to reach maturity and produced up to 12% lower yields than their non-GM counterparts.[3] And trials of canola varieties in Australia conducted by the Grains Research and Development Council found that yields were 0.7 tonnes per hectare for GM and 0.8 tonnes per hectare for non-GM.[4]

In 2009, in an apparent attempt to counter criticisms of lower yields from its GM soy, Monsanto released its new generation of supposedly high-yielding GM soybeans, RR2 Yield®. But a study carried out in five US states involving farm managers who planted RR2 soybeans concluded that the new varieties "didn't meet their [yield] expectations".[5]

A US Department of Agriculture (USDA) report confirmed in 2014, "Over the first 15 years of commercial use, GE seeds have not been shown to increase yield potentials of the varieties. In fact, the yields of herbicide-tolerant [HT] or insect-resistant seeds may be occasionally lower than the yields of conventional varieties if the varieties used to carry the HT or Bt genes are not the highest yielding cultivars, as in the earlier years of adoption."[6]

These findings should not surprise us. Yield is a complex trait that is the product of many genes working together – it cannot be genetically engineered into a crop. Virtually all commercially available GM crops are engineered to make the plants tolerate herbicides or express an insecticidal protein. A high-yielding GM crop is a conventional breeding success into which herbicide tolerance and/or insect-resistance genes have been added through genetic engineering.

Klümper and Qaim meta-analysis

A 2014 meta-analysis by Klümper and Qaim claimed that GM crops have "increased crop yields by 22%, and increased farmer profits by 68%".[7] But the authors' yield figures included suspect data on GM Bt cotton collected from Monsanto field trials and published in an earlier paper by Qaim,[8] which drew intense criticism for claiming 80% yield hikes for Bt over non-Bt cotton.[9,10] Even the former Syngenta employee and GMO proponent Dr Shanthu Shantharam attacked the earlier paper as a "shoddy publication based on meagre and questionable field data".[10]

Almost all time periods measured were prior to the rise of herbicide-resistant weeds and Bt toxin-resistant pests, making the claimed benefits a short-term phenomenon. The meta-analysis also does not reflect the true balance of GM crops grown worldwide. Nearly 80% of the studies are on Bt insecticidal crops, despite the fact that over 80% of GM crops grown worldwide are herbicide-tolerant. And over 50% of the studies are just on Bt cotton. Only 14% of the studies were on herbicide-tolerant soybeans, despite this combination of crop and trait being the most widely planted and longest in commercialization.[11]

The meta-analysis is heavily weighted towards data from short-term and small-scale trials early in the release period for any specific GMO. Thus it confuses the well-known biasing factor of good farmers devoting extra resources to expensive seeds[12] with their true performance in large-scale cultivation.

The more relevant comparison would be large-scale long-term analyses of yields in real farming conditions, such as that conducted by Heinemann and colleagues. This peer-reviewed study analyzed data on agricultural productivity in the United States and Western Europe over the last 50 years, focusing on maize, canola, and wheat. While the study was not about GM, it found yield declines in the US's largely GMO cropping systems, compared with Western Europe's largely non-GMO systems. In addition, Western Europe's yield gains were achieved with less pesticide use than in the US.[13,14]

Failure to yield

The report "Failure to yield", by Dr Doug Gurian-Sherman, is an important contribution to our understanding of the yield performance of GM crops. At the time he wrote the report, Dr Gurian-Sherman was senior scientist at the Union of Concerned Scientists. He is a former biotechnology adviser to the US Environmental Protection Agency and is now at the Center for Food Safety (USA).

The report, based on peer-reviewed research and official USDA data, was the first to tease out the contribution of genetic engineering to yield performance from the gains made through conventional breeding.[15] This is important because GMO companies insert their proprietary GM genes into the best-performing conventionally bred varieties.

The report also differentiated between intrinsic and operational yield.[15] Intrinsic or potential yield is the highest yield that can be achieved when crops are grown under ideal conditions. Operational yield is what the farmer ends up with once environmental factors such as pests and bad weather conditions have taken their toll on yields. Genes that improve operational yield can reduce losses from such factors. For example, a Bt trait that kills an insect pest could save the crop from some yield loss, but only in years when that pest is a significant problem.

The study found that GM technology has not raised the intrinsic yield of any crop. The intrinsic yields of maize and soybeans rose during the twentieth century, but this was not as a result of GM traits, but due to improvements from traditional breeding.[15]

Looking at individual crops, GM soybeans have not increased operational yields either. GM Bt maize increased operational yields slightly in years of heavy infestation with the European corn borer pest. GM Bt maize offered little or no advantage when infestation with European corn borer was low to moderate, even compared with non-GM maize untreated with insecticides.[15]

"Failure to yield" concluded, "Commercial GE crops have made no inroads so far into raising the intrinsic or potential yield of any crop. By contrast, traditional breeding has been spectacularly successful in this regard; it can be solely credited with the intrinsic yield increases in the United States and other parts of the world that characterized the agriculture of the twentieth century."[15]

Why do some farmers believe that GM crops have better yields than non-GM crops?

In countries where GM seeds dominate the market, and in crops that are intensively bred for yield, a trend has grown among GMO seed companies of neglecting the development of non-GMO varieties. The result is that the best naturally high-yielding crop varieties are only available with GMO traits added. The non-GMO growers are left with years-old germplasm that yields less well than the latest improved varieties – with added GMO traits.[16] This phenomenon does not mean that GM crops have better yields. It means that non-GMO breeding has been neglected in favour of developing GM crops that are more easily patented and bring more profit to the big GMO seed companies.

Bt cotton in India

Since the introduction of GM Bt cotton into India, there has been a fierce debate about its performance. Some claim it has delivered higher yields and improved farmer income, while others claim it has suffered widespread failure and led to farmer suicides. The main difficulty in resolving the argument is that good comparative data on Bt and non-Bt cotton performance do not exist. Typically such data would be generated from a study in which two groups of farmers matched for ability and farm conditions would be assigned to grow Bt cotton or non-Bt cotton.

Some of the most nuanced analyses of the performance of Bt cotton are by the US academic Glenn Davis Stone, who mistrusts "narratives" regarding Bt cotton from both sides of the debate. According to Stone, within five years of the introduction of Bt cotton in India, national cotton yields rose by 84%. However, almost all of that rise occurred in 2003/4, when only 1.2% of the cotton was Bt, and 2004/5, when only 5.6% of the cotton was Bt. In short, Stone concluded, "Bt couldn't have been responsible for the rise."[17]

What is more, wrote Stone in 2012, "In the last four years, as Bt has

All India cotton yields and Bt percentage of cotton area

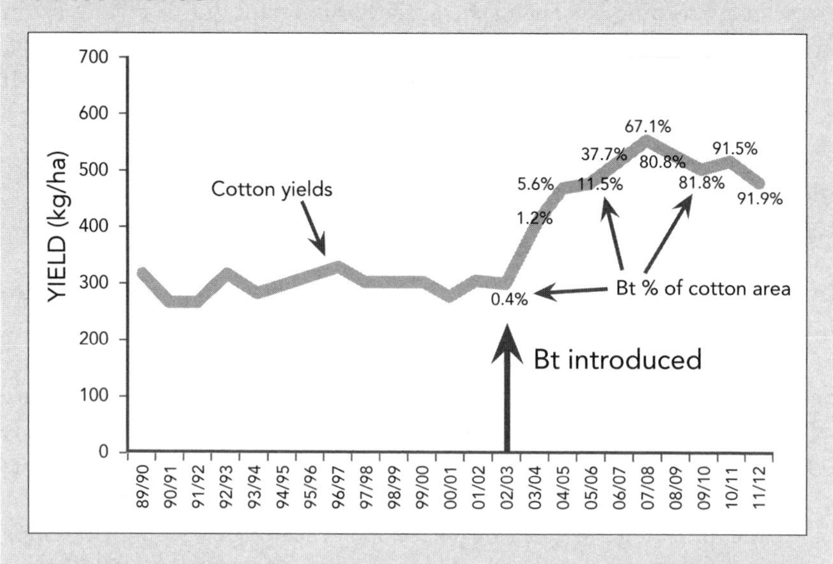

India approved Bt cotton in 2002; in 2012 it accounted for 92% of all Indian cotton. Average nationwide cotton yields went from 302 kg/ha in the 2002/3 season to a projected 481 kg/ha in 2011/12 – up 59.3% overall. This chart shows the trends in yields, which took off after Bt was introduced in 2002.

The problem is that while yields did take off right after Bt cotton was approved, this was well before Bt cotton was widely adopted. This graph shows the yearly percentages of all Indian cotton land planted to Bt cotton. Most of the yield increase happened between 2002–2005, when Bt only comprised between 0.4-5.6% of India's cotton. Obviously Bt couldn't have accounted for more than a tiny speck of the national rise.

Charts and data adapted from Glenn Davis Stone, "Bt cotton, remarkable success, and four ugly facts", fieldquestions, 12 February 2012. http://fieldquestions.com/2012/02/12/bt-cotton-remarkable-success-and-four-ugly-facts/

Yield data is from India's Cotton Advisory Board, downloaded 28 Jan 2012 from the website of the Cotton Corporation of India Ltd (http://cotcorp.gov.in/state-operations.aspx): "State-wise area, production, yield for last ten years". Bt cotton adoption data is from ISAAA.

risen from 67% to 92% of India's cotton, yields have dropped steadily."[17] According to India's Cotton Advisory Board, yield declined to a five-year low in 2012–13.[18]

Stone summed up the situation as follows: "Bt didn't explain the big rise in yields, and since Bt has taken over, yields have been steadily worsening."[19]

So what explained the temporary rise in yields, if it wasn't Bt cotton? Dr Keshav Kranthi, director of India's Central Institute for Cotton Research, believes that the other factor responsible for the short-term yield gains was the use of insecticides against sap-sucking pests such as the leaf hopper.

Dr Kranthi wrote, "Since 2002, every Bt cotton seed has been treated with the highly effective insecticide, imidacloprid." This insecticide also could also have accounted for the recent decline in cotton yield, since "Recently, leaf hoppers were found to have developed resistance to imidacloprid and... yields are likely to decline."[20]

Research from the University of Hannover, Germany, in cooperation with the Food and Agriculture Organization of the United Nations, found that Bt cotton performed better under irrigated conditions but that non-Bt local varieties were better suited to rainfed conditions. However, the yield advantage of Bt cotton under irrigation was offset by higher production costs and lower product prices.[21]

References

1. Benbrook C. Evidence of the Magnitude and Consequences of the Roundup Ready Soybean Yield Drag from University-Based Varietal Trials in 1998: Ag BioTech InfoNet Technical Paper Number 1. Sandpoint, Idaho; 1999. http://www.mindfully.org/GE/RRS-Yield-Drag.htm.
2. Elmore RW, Roeth FW, Nelson LA, et al. Glyphosate-resistant soyabean cultivar yields compared with sister lines. Agron J. 2001;93:408-412.
3. Ma BL, Subedi KD. Development, yield, grain moisture and nitrogen uptake of Bt corn hybrids and their conventional near-isolines. Field Crops Res. 2005;93:199-211.
4. Bennett H. GM canola trials come a cropper. WA Business News. http://www.wabusinessnews.com.au/en-story/1/69680/GM-canola-trials-come-a-cropper. Published January 16, 2009.
5. Kaskey J. Monsanto facing "distrust" as it seeks to stop DuPont (update 3). Bloomberg. http://www.bloomberg.com/apps/news?pid=newsarchive&sid=aii_24MDZ8SU. Published November 10, 2009.
6. Fernandez-Cornejo J, Wechsler S, Livingston M, Mitchell L. Genetically Engineered Crops in the United States. Washington, DC: US Department of Agriculture; 2014. http://www.ers.usda.gov/publications/err-economic-research-report/err162.aspx#.U0P_qMfc26x.
7. Klümper W, Qaim M. A meta-analysis of the impacts of genetically modified crops. PLoS ONE. 2014;9(11):e111629. doi:10.1371/journal.pone.0111629.
8. Qaim M, Zilberman D. Yield effects of genetically modified crops in developing countries. Science. 2003;299(5608):900-902. doi:10.1126/science.1080609.
9. Sharma D. A scientific fairytale. India Together. http://indiatogether.org/scicoverup-op-ed. Published February 1, 2003.
10. Shantharam S. Outstanding performance of Bt cotton in India... really? AgBioIndia Bulletin. http://www.lobbywatch.org/archive2.asp?arcid=85. Published February 26, 2003.
11. Heinemann J. Correlation is not causation. RightBiotech. http://rightbiotech.tumblr.com/post/103665842150/correlation-is-not-causation. Published November 27, 2014.
12. Stone GD. Constructing facts: Bt cotton narratives in India. Econ Polit Wkly. 2012;47(38):62-70.

13. Heinemann JA, Massaro M, Coray DS, Agapito-Tenfen SZ, Wen JD. Sustainability and innovation in staple crop production in the US Midwest. Int J Agric Sustain. 2013:1-18.
14. Heinemann JA, Massaro M, Coray DS, Agapito-Tenfen SZ. Reply to comment on sustainability and innovation in staple crop production in the US Midwest. Int J Agric Sustain. 2014:1-4. doi:10.1080/14735903.2014.939843.
15. Gurian-Sherman D. Failure to Yield: Evaluating the Performance of Genetically Engineered Crops. Cambridge, MA: Union of Concerned Scientists; 2009. http://www.ucsusa.org/assets/documents/food_and_agriculture/failure-to-yield.pdf.
16. Roseboro K. Farmers' seed options drastically reduced in GMO-producing countries. The Organic and Non-GMO Report. http://www.non-gmoreport.com/articles/march2013/farmers-seed-options-GMO-producing-countries.php. Published February 28, 2013.
17. Stone GD. Bt cotton, remarkable success, and four ugly facts. FieldQuestions.com. http://fieldquestions.com/2012/02/12/bt-cotton-remarkable-success-and-four-ugly-facts/. Published February 12, 2012.
18. Jha DK. Bt cotton losing steam, productivity at 5-yr low. Business Standard (India). http://www.business-standard.com/article/markets/bt-cotton-losing-steam-productivity-at-5-yr-low-113020601016_1.html. Published February 7, 2013.
19. Stone GD. Bt cotton is failing: Blame the farmers. FieldQuestions.com. 2013. http://fieldquestions.com/2013/02/09/bt-cotton-is-failing-blame-the-farmers/.
20. Kranthi KR. Part II: 10 Years of Bt in India. Cotton Grower. http://www.cottongrower.com/uncategorized/part-ii-10-years-of-bt-in-india/. Published May 1, 2011.
21. Orphal J. Comparative Analysis of the Economics of Bt and Non-Bt Cotton Production. Hannover, Germany: University of Hannover, Germany, with Food and Agriculture Organization of the United Nations; 2005. http://www.ifgb.uni-hannover.de/fileadmin/eagr/EUE_files/PPP_Publicat/Special_Series/ppp_s08.pdf.

9 **Myth:** GM crops decrease pesticide use

Truth: GM crops increase pesticide use

Myth at a glance

GM crops are claimed by proponents to reduce pesticide use (the term "pesticide" includes herbicides, which technically are pesticides). But this is untrue.

Since GM crops were introduced in the US, overall pesticide use has increased by an estimated 404 million pounds (183 million kg), or about 7%, compared with what would have been used if the same acres had been planted to non-GM crops.

The small reduction in the use of chemical insecticide sprays due to GM Bt insecticidal crops is swamped by the large increase in herbicide use due to GM herbicide-tolerant crops.

The widespread use of herbicide-tolerant crops has led to the rapid spread of herbicide-resistant superweeds. The area of US cropland infested with glyphosate-resistant weeds expanded to 61.2 million acres in 2012. In some areas, farmland has been abandoned or farmers have resorted to pulling weeds by hand.

The GMO industry's answer to glyphosate-resistant superweeds has been to develop crops that tolerate other, potentially even more toxic herbicides and mixtures of herbicides. In 2014 the US Department of Agriculture approved Dow's maize and soybean varieties engineered to tolerate application of the herbicide 2,4-D as well as glyphosate.

This "chemical treadmill" model of farming is unsustainable and especially impractical for farmers in the Global South, who cannot afford expensive pesticides.

GM Bt crops are in themselves an insecticide, so overall they do not eliminate or reduce insecticides, but change the type of insecticide and the way in which it is used – from sprayed on, to built-in.

As insect pests are becoming resistant to the Bt toxins in GM Bt crops, farmers are turning back to chemical insecticides.

In addition, the use of toxic neonicotinoid insecticidal seed treatments steeply increased during the 2000s on GM and non-GM maize and soy alike. These seed treatments have not been considered in studies claiming reductions in insecticide use from GM Bt crops. Thus such claims are invalid.

Over 99% of all commercialized GM crops are engineered to tolerate one or more herbicides, or to express an insecticide, or both.[1] The most widely grown GM crop is Roundup Ready (RR) soy,[2] engineered to tolerate the glyphosate-based herbicide Roundup. The RR gene in GM crops enables farmers to spray the field liberally with Roundup herbicide. All plant life is killed except the crop.

The widespread adoption of GM RR soy in North and South America has led to large increases in the use of Roundup and other glyphosate-based herbicides.[3,4,5,6,7,8] An analysis of US Department of Agriculture data found that GM herbicide-tolerant crops have led to a 239 million kilogram (527 million pound) increase in herbicide use in the US in the first 16 years of cultivation, while Bt crops reduced chemical insecticide spray use by 56 million kilograms (123 million pounds). Overall pesticide use increased by an estimated 183 million kg (404 million pounds), or about 7%, compared with the amount that would have been used if the same acres had been planted to non-GM crops.[4]

A 2014 meta-analysis by Klümper and Qaim, which has been much cited by GMO proponents, claimed that GM crops have "reduced chemical pesticide use by 37%... and increased farmer profits by 68%".[9] But it relied heavily on outdated data from the early 2000s – before herbicide-resistant superweeds and Bt resistant insect pests made GM herbicide-tolerant and Bt insecticidal traits less effective (see below), impacting farmers' yields and profits. Also, important parts of the data used were generated in Monsanto-funded Bt cotton field trials and their relevance to real farm conditions is questionable.

GM crops have created "superweeds"

The major cause of the increase in herbicide use in GM crop fields is the rapid spread of glyphosate-resistant superweeds.[4] When resistant weeds first appear, farmers use more glyphosate herbicide to try to control them. But as time passes, no amount of glyphosate herbicide is effective.[10,11] Farmers resort to potentially even more toxic herbicides and mixtures of herbicides in attempts to control weeds, including 2,4-D (an ingredient of the Vietnam War toxic defoliant Agent Orange) and dicamba.[3,12,13,14,15,16,17,18]

Some US farmers are going back to more labour-intensive methods like ploughing – and even pulling weeds by hand.[19] In Georgia in 2007, 10,000 acres of farmland were abandoned after being overrun by glyphosate-resistant pigweed.[20] Resistant pigweed in the Southern United States was so tough that it broke farm machinery.[21] An article in the farm press in 2015 reported that hundreds of thousands of acres of farmland in the Southeast United States was being converted from conservation tillage, where weeds are controlled with herbicides instead of ploughing, back to ploughing, in an attempt to conquer herbicide-resistant weeds.[22]

A maize field in Georgia, USA, in 2010, over-run with glyphosate-resistant palmer amaranth

The GM industry "solution" to superweeds: More herbicides

The industry's answer to glyphosate-tolerant superweeds has been to develop stacked-trait GM crop varieties that are resistant to multiple herbicides. In 2014 the US Department of Agriculture approved Dow's multi-herbicide-tolerant GM soybean, engineered to tolerate being sprayed with glyphosate, glufosinate, and 2,4-D,[23] and a 2,4-D-tolerant maize.[24] Weed scientists warn that such multi-herbicide-tolerant crops will cause an increase in 2,4-D use, trigger an outbreak of still more intractable weeds resistant to both glyphosate and 2,4-D, and undermine sustainable approaches to weed management.[3] In 2015 the US Department of Agriculture approved Monsanto's GM soybeans and cotton, engineered to tolerate the herbicide dicamba as well as glyphosate.[25] The USDA's Environmental Impact Statement predicted that dicamba use will increase 88-fold and 14-fold for soybeans and cotton respectively, compared to current levels.[26]

Weed species already exist that are resistant to dicamba,[27] 2,4-D,[28] and multiple herbicides.[29]

Keeping farmers tied to a chemical treadmill benefits GMO seed companies, which are also agrochemical companies.[30,31]

Herbicide-tolerant crops undermine sustainable agriculture

"Agricultural weed management has become entrenched in a single tactic – herbicide-resistant crops – and needs greater emphasis on integrated practices that are sustainable over the long term. In response to the outbreak of glyphosate-resistant weeds, the seed and agrichemical industries are developing crops that are genetically modified to have combined resistance to glyphosate and synthetic auxin [plant hormone] herbicides. This technology will allow these herbicides to be used over vastly expanded areas and will likely create three interrelated challenges for sustainable weed management. First, crops with stacked herbicide resistance are likely to increase the severity of resistant weeds. Second, these crops will facilitate a significant increase in herbicide use, with potential negative consequences for environmental quality. Finally, the short-term fix provided by the new traits will encourage continued neglect of public research and extension in integrated weed management."

– David A. Mortensen, professor of weed and applied plant ecology, Penn State University, and colleagues[3]

GM Bt crops do not reduce insecticides but change the way they're used

GM proponents claim that GM Bt crops reduce insecticide use, as farmers do not have to spray chemical insecticides. But this claim does not stand up to analysis, because GM Bt crops are in themselves insecticides. The GM insecticide is present in active form in every part of the crop. So Bt crops do not reduce or eliminate insecticides. They simply change the type of insecticide and the way in which it is used – from sprayed on, to built-in. Claims that GM Bt crops reduce or eliminate insecticides fail to take these plant-produced pesticides into account.

What is more, the amount of Bt insecticide produced by GM Bt crops is in most cases far greater than the amount of chemical insecticide spray that these crops displace.[4] For example, GM Bt maize DAS 59122–7, targeting the corn rootworm, expresses two Bt toxin proteins totalling 2.8 kg/ha, 14-fold more than the chemical insecticides displaced.[4]

SmartStax GM maize produces six Bt toxin proteins targeting the European corn borer and the corn rootworm. Total Bt toxin protein production is estimated at 4.2 kg/ha, 19 times the average chemical insecticide rate of application on non-GM maize in 2010.[4] This high level of Bt toxin expressed in SmartStax has never been tested in animals or humans to see if it is safe to eat over the long term.

The large quantities of Bt toxins in GM crops are seen as acceptable by proponents of GMO foods as Bt toxin is claimed as being non-hazardous to the environment and consumers. However, the Bt toxins engineered into GM crops are different from natural Bt toxins. They are not harmless, as GMO proponents claim, but have been found to have harmful effects on butterflies and other non-target and beneficial insects (see Chapter 6), as well as mammals (see Chapter 4).

Major weakness of studies claiming reduced insecticide use from GM Bt crops

A study[32] by Margaret Douglas and John Tooker from Pennsylvania State University pointed out a major weakness in Klümper and Qaim's meta-analysis, which claimed a 42% reduction in pesticide use due to GM Bt insecticidal crops.[9] Klümper and Qaim, in common with other authors who claim reduced insecticide use from GM Bt crops,[33] failed to take into account the escalating use of neonicotinoid insecticidal seed treatments in major crops.

Douglas and Tooker found that neonicotinoid use increased rapidly in the US between 2003 and 2011, with 79–100% of maize and 34–44% of soybean hectares being treated in 2011.[32]

Douglas and Tooker explained that studies such as Klümper and Qaim's "do not seem to have considered seed treatments, and so may have overstated reductions in insecticide use".[32]

Neonicotinoid seed treatments are applied to both GM and non-GM crops. However, Douglas and Tooker's study period[32] coincided with the expansion of the GM maize and soybeans in the US[34] and they note that these two crops have driven the major part of the increase in neonicotinoid use.[32] Their study exposes claims of insecticide reductions due to GM Bt crops as baseless.

GM Bt crop farmers don't always give up chemical insecticide sprays

GMO proponents often assume that farmers who adopt Bt crops give up chemical insecticides – but studies have shown this to be a faulty assumption. Tabashnik (2008) reported that while bollworms have evolved resistance to Bt toxin in GM cotton, this has not caused widespread crop failure because "insecticides have been used from the outset" to control the pest.[35] A Greenpeace study comparing the livelihoods of Bt cotton and organic cotton farmers in India found that Bt cotton farmers continued to use a large amount of chemical insecticides, including some classified by the World Health Organization as extremely or highly hazardous.[36]

Claims of reductions in insecticide use from Bt crop adoption are unreliable unless there is specific evidence that the farmer does not use chemical insecticides and insecticidal seed treatments are not used.

Resistant pests are making GM Bt technology obsolete

GM Bt insecticidal crops express the Bt toxin in every cell for their entire lifetime, constantly exposing pests to the toxin. This is a recipe for rapid evolution of resistance, since only the most resistant pests survive exposure to the toxin, reproduce, and pass on their resistance genes.

For this reason, GM Bt crop technology sometimes enjoys short-term success in controlling pests but is soon undermined by the emergence of pests resistant to the toxin.[35,37,38] By 2011 corn rootworms in some areas of the US were already resistant to two of the three available Bt toxins that previously controlled them, resulting in severe crop damage.[39] Bt-resistant rootworm populations have been reported in Iowa[40,41] and Illinois.[42,43] In Brazil, pests became resistant to the Bt toxin in GM maize 1507 within three years of the maize first being grown.[44]

Entomologist Elson Shields of Cornell University commented on the evolution of pest resistance to Bt toxin in GM plants, "The insect will win. Always bet on the insect if there is not a smart deployment of the trait."[39]

Secondary pests move in on GM Bt crops

Even when Bt toxin succeeds in controlling the target pest, secondary pests can move into the ecological niche. In the US, the Western bean cutworm has increased significantly in GM Bt maize fields.[45] In India and China, Bt cotton was initially effective in suppressing the target pest, the bollworm. But secondary pests that are resistant to Bt toxin, especially mirids and mealy bugs, soon took its place.[46,47,48,49,50,51] Studies from China show that GM Bt cotton is already failing under the onslaught of secondary pests.[52,53]

GM farming systems lag behind non-GM in producing higher yields with less pesticide

A peer-reviewed study analyzed data on agricultural productivity in the United States and Western Europe over the last 50 years, focusing on maize, canola, and wheat. The study found that the US's largely GM production has lower yields and higher pesticide use compared to Western Europe's largely non-GM systems.[54]

The study found that both herbicide and insecticide use is increasing in the US relative to Western Europe. So the mostly non-GMO farming system of Western Europe is reducing chemical inputs and thus becoming more sustainable than that of the US, without sacrificing yield gains.

Author Professor Jack Heinemann commented, "The US and US industry have been crowing about the reduction in chemical insecticide use with the introduction of Bt [GM insecticidal] crops. And at face value, that's true. They've gone to about 85% of the levels that they used in the pre-GE era. But what they don't tell you is that France went down to 12% of its previous levels. France is the fourth biggest exporter of corn in the world, one of the biggest exporters of wheat, and it's only 11% of the size of the US.[55]

Heinemann was prompted to carry out the study by a claim from a British economics professor that Europe was falling behind the US in agricultural productivity because of its avoidance of GM. Heinemann's team found that the opposite is true: "Europe has learned to grow more food per hectare and use fewer chemicals in the process. The American choices in biotechnology [GM among them] are causing it to fall behind Europe in productivity and sustainability."[56]

References

1. International Service for the Acquisition of Agri-biotech Applications (ISAAA). ISAAA Brief 49-2014: Executive Summary: Global Status of Commercialized biotech/GM Crops: 2014. Ithaca, NY: International Service for the Acquisition of Agri-biotech Applications (ISAAA); 2014. http://www.isaaa.org/resources/publications/briefs/49/executivesummary/default.asp.
2. James C. Global Status of Commercialized biotech/GM Crops: 2012. Ithaca, NY: ISAAA; 2012. http://www.isaaa.org/resources/publications/briefs/44/download/isaaa-brief-44-2012.pdf.
3. Mortensen DA, Egan JF, Maxwell BD, Ryan MR, Smith RG. Navigating a critical juncture for sustainable weed management. BioScience. 2012;62(1):75-84.
4. Benbrook C. Impacts of genetically engineered crops on pesticide use in the US – The first sixteen years. Environ Sci Eur. 2012;24(24). doi:10.1186/2190-4715-24-24.

5. Benbrook CM. Rust, Resistance, Run down Soils, and Rising Costs – Problems Facing Soybean Producers in Argentina. Technical Paper No 8. AgBioTech InfoNet; 2005. http://www.greenpeace.org/raw/content/international/press/reports/rust-resistence-run-down-soi.pdf.
6. Pengue W. El glifosato y la dominación del ambiente. Biodiversidad. 2003;37. http://www.grain.org/biodiversidad/?id=208.
7. MECON (Ministerio de Economia Argentina). Mercado argentino de fitosanitarios – Año 2001. 2001. http://bit.ly/1eqMudL.
8. CASAFE. Mercado Argentino de productos fitosanitarios 2012. 2012. http://www.casafe.org/pdf/estadisticas/Informe%20Mercado%20Fitosanitario%202012.pdf.
9. Klümper W, Qaim M. A meta-analysis of the impacts of genetically modified crops. PLoS ONE. 2014;9(11):e111629. doi:10.1371/journal.pone.0111629.
10. Nandula VK, Reddy KN, Duke SO, Poston DH. Glyphosate-resistant weeds: Current status and future outlook. Outlooks Pest Manag. 2005;16:183-187.
11. Syngenta. Syngenta module helps manage glyphosate-resistant weeds. Delta Farm Press. http://deltafarmpress.com/syngenta-module-helps-manage-glyphosate-resistant-weeds. Published May 30, 2008.
12. Robinson R. Resistant ryegrass populations rise in Mississippi. Delta Farm Press. 2008. http://deltafarmpress.com/resistant-ryegrass-populations-rise-mississippi.
13. Johnson B, Davis V. Glyphosate resistant horseweed (marestail) found in 9 more Indiana counties. Pest Crop. 2005. http://extension.entm.purdue.edu/pestcrop/2005/issue8/index.html.
14. Nice G, Johnson B, Bauman T. A little burndown madness. Pest & Crop. http://extension.entm.purdue.edu/pestcrop/2008/issue1/index.html. Published March 7, 2008.
15. Nice G, Johnson B. Fall applied programs labeled in Indiana. Pest Crop. 2006;(23). http://extension.entm.purdue.edu/pestcrop/2006/issue23/table1.html.
16. Randerson J. Genetically-modified superweeds "not uncommon." New Sci. 2002. http://www.newscientist.com/article/dn1882-geneticallymodified-superweeds-not-uncommon.html.
17. Kilman S. Superweed outbreak triggers arms race. Wall Street Journal. http://biolargo.blogspot.com/2010/06/round-up-weed-killer-and-acquired.html. Published June 4, 2010.
18. Brasher P. Monsanto paying farmers to increase herbicide use. Des Moines Register. http://bit.ly/az3fSo. Published October 19, 2010.
19. Neuman W, Pollack A. US farmers cope with Roundup-resistant weeds. New York Times. http://www.nytimes.com/2010/05/04/business/energy-environment/04weed.html?pagewanted=1&hp. Published May 3, 2010.
20. Caulcutt C. "Superweed" explosion threatens Monsanto heartlands. France 24. http://www.gmwatch.org/index.php/news/archive/2009/10923. Published April 19, 2009.
21. Osunsami S. Killer pig weeds threaten crops in the South. http://abcnews.go.com/WN/pig-weed-threatens-agricultureindustryovertaking-fields-crops/story?id=8766404&page=1. Published October 6, 2009.
22. Hollis P. Conservation tillage systems threatened by herbicide-resistant weeds. Southeast Farm Press. http://southeastfarmpress.com/management/conservation-tillage-systems-threatened-herbicide-resistant-weeds?page=1. Published March 11, 2015.
23. Gillam C. Dow launches multi-herbicide tolerant GM soybean. Reuters. http://bit.ly/qBR9a5. Published August 22, 2011.
24. Kimbrell A. "Agent Orange" corn: Biotech only winner in chemical arms race as herbicide resistant crops fail. Huffington Post. http://www.huffingtonpost.com/andrew-kimbrell/agent-orange-corn-biotech_b_1291295.html. Published February 22, 2012.
25. Food & Water Watch. USDA says "yes" to pesticide drift, approves dicamba-tolerant crops. Food & Water Watch. http://www.foodandwaterwatch.org/pressreleases/usda-says-yes-to-pesticide-drift-approves-dicamba-tolerant-crops%E2%80%A8/. Published January 15, 2015.
26. US Department of Agriculture (USDA). Monsanto Petitions (10-188-01p and 12-185-01p) for Determinations of Nonregulated Status for Dicamba-Resistant Soybean and Cotton Varieties: Final Environmental Impact Statement. Riverdale, MD: US Department of Agriculture; 2014. http://www.aphis.usda.gov/brs/aphisdocs/dicamba_feis.pdf.
27. Rahman A, James TK, Trolove MR. Chemical control options for the dicamba resistant biotype of fathen (Chenopodium album). N Z Plant Prot. 2008;61:287-291.
28. Heap I. International Survey of Herbicide Resistant Weeds: Weeds resistant to synthetic auxins (O/4) by species and country. 2014. http://weedscience.org/Summary/MOA.aspx?MOAID=24.
29. Martin H. Herbicide Resistant Weeds. Ontario Ministry of Agriculture, Food and Rural Affairs; 2013. http://www.omafra.gov.on.ca/english/crops/facts/01-023.htm.
30. Howard P. Visualizing consolidation in the global seed industry: 1996–2008. Sustainability. 2009;1:1266-1287.
31. Howard P. Seed industry structure 1996–2013. Philip H Howard Assoc Profr Mich State Univ. 2014. https://www.msu.edu/~howardp/seedindustry.html.

32. Douglas MR, Tooker JF. Large-scale deployment of seed treatments has driven rapid increase in use of neonicotinoid insecticides and preemptive pest management in U.S. field crops. Environ Sci Technol. 2015. doi:10.1021/es506141g.
33. Naranjo SE. Impacts of Bt crops on non-target invertebrates and insecticide use patterns. CAB Rev Perspect Agric Vet Sci Nutr Nat Resour. 2009;4(011).
34. US Department of Agriculture (USDA) Economic Research Service. Recent trends in GE adoption. 2014. http://www.ers.usda.gov/data-products/adoption-of-genetically-engineered-crops-in-the-us/recent-trends-in-ge-adoption.aspx.
35. Tabashnik BE, Gassmann AJ, Crowder DW, Carriere Y. Insect resistance to Bt crops: Evidence versus theory. Nat Biotechnol. 2008;26:199-202. doi:10.1038/nbt1382.
36. Tirado R. Picking Cotton: The Choice between Organic and Genetically Engineered Cotton for Farmers in South India. Amsterdam, The Netherlands: Greenpeace; 2010. http://www.greenpeace.org/international/Global/international/publications/agriculture/2010/Picking_Cotton.pdf.
37. Rensburg JBJ. First report of field resistance by the stem borer, Busseola fusca (Fuller) to Bt-transgenic maize. Afr J Plant Soil. 2007;24:147-151.
38. Huang F, Leonard BR, Wu X. Resistance of sugarcane borer to Bacillus thuringiensis Cry1Ab toxin. Entomol Exp Appl. 2007;124:117-123.
39. Keim B. Voracious worm evolves to eat biotech corn engineered to kill it. Wired.com. 2014. http://www.wired.com/2014/03/rootworm-resistance-bt-corn/.
40. Gassmann AJ, Petzold-Maxwell JL, Keweshan RS, Dunbar MW. Field-evolved resistance to Bt maize by Western corn rootworm. PLoS ONE. 2011;6:e22629. doi:10.1371/journal.pone.0022629.
41. Associated Press. Monsanto shares slip on bug-resistant corn woes. Bloomberg Businessweek. http://www.businessweek.com/ap/financialnews/D9PDS5KO0.htm. Published August 29, 2011.
42. Gray M. Western corn rootworm resistance to Bt corn confirmed in more Illinois counties. Aces News. http://www.aces.uiuc.edu/news/stories/news5903.html. Published April 7, 2014.
43. Gillam C. GMO corn failing to protect fields from pest damage – report. Reuters. http://www.reuters.com/article/2013/08/28/usa-gmo-corn-rootworm-idUSL2N0GT1ED20130828. Published August 28, 2013.
44. Farias JR, Andow DA, Horikoshi RJ, et al. Field-evolved resistance to Cry1F maize by Spodoptera frugiperda (Lepidoptera: Noctuidae) in Brazil. Crop Prot. 2014;64:150-158. doi:10.1016/j.cropro.2014.06.019.
45. Dorhout DL, Rice ME. Intraguild competition and enhanced survival of western bean cutworm (Lepidoptera: Noctuidae) on transgenic Cry1Ab (MON810) Bacillus thuringiensis corn. J Econ Entomol. 2010;103:54-62.
46. Pearson H. Transgenic cotton drives insect boom. Nature. 2006. doi:10.1038/news060724-5.
47. Wang S, Just DR, Pinstrup-Andersen P. Bt-cotton and secondary pests. Int J Biotechnol. 2008;10:113-121.
48. Goswami B. Making a meal of Bt cotton. Infochange. 2007. http://infochangeindia.org/other/features/making-a-meal-of-bt-cotton.html?Itemid=.
49. Ashk GKS. Bt cotton not pest resistant. The Times of India. http://timesofindia.indiatimes.com/Chandigarh/Bt_cotton_not_pest_resistant/articleshow/2305806.cms. Published August 24, 2007.
50. The Economic Times (India). Bug makes meal of Punjab cotton, whither Bt magic?http://www.gmwatch.org/latest-listing/46-2007/7640. Published September 2, 2007.
51. Rohini RS, Mallapur CP, Udikeri SS. Incidence of mirid bug, Creontiades biseratense (Distant) on Bt cotton in Karnataka. Karnataka J Agric Sci. 2009;22:680-681.
52. Zhao JH, Ho P, Azadi H. Benefits of Bt cotton counterbalanced by secondary pests? Perceptions of ecological change in China. Env Monit Assess. 2010;173:985-994. doi:10.1007/s10661-010-1439-y.
53. Lu Y, Wu K, Jiang Y, et al. Mirid bug outbreaks in multiple crops correlated with wide-scale adoption of Bt cotton in China. Science. 2010;328:1151-1154. doi:10.1126/science.1187881.
54. Heinemann JA, Massaro M, Coray DS, Agapito-Tenfen SZ, Wen JD. Sustainability and innovation in staple crop production in the US Midwest. Int J Agric Sustain. 2013:1-18.
55. Richardson J. Study: Monsanto GMO food claims probably false. Salon.com. 2013. http://www.salon.com/2013/06/27/study_monsanto_gmo_food_claims_probably_false_partner/.
56. University of Canterbury. GM a failing biotechnology in modern agro-ecosystems [press release]. 2013. http://www.gmwatch.org/index.php/news/rss/14802.

10 **Myth:** The pesticides associated with GM crops are safe

Truth: The pesticides associated with GM crops may pose risks to health and the environment

Myth at a glance

GM herbicide-tolerant crops absorb herbicides into their tissues. People and animals that eat these crops are eating herbicide residues.

Studies show that the herbicides used on GM crops – Roundup (based on the chemical glyphosate), glufosinate, 2,4-D, and dicamba – are toxic to humans and animals.

In 2015 the World Health Organization's cancer agency declared that Roundup, used on over 80% of GM crops worldwide, "probably" causes cancer. Levels of glyphosate found in human urine are similar to levels found to stimulate human breast cancer cell growth in cell culture.

Safety tests of pesticides conducted for regulatory approvals are based on the old notion that toxicity always increases with dose and the low levels we are exposed to are safe. This is out of step with research that shows that pesticides can cause endocrine (hormone) disruption and harm to health at low doses that regulators classify as safe.

In addition to the primary active ingredient (glyphosate in the case of Roundup), pesticides contain additives or "adjuvants", which are not well tested for safety. Also, the complete pesticide formulations as sold and used are not tested for long-term safety for regulatory approvals. Only the stated "active ingredient" is tested and these limited tests are used to set regulatory safety limits. Yet formulations have generally been found to be more toxic than isolated active ingredients, due to the adjuvants. Thus so-called safe levels may not be safe.

The Bt toxins expressed in GM Bt insecticidal crops are different from natural Bt sprayed by organic and conventional farmers. GM Bt crops cannot be assumed to be safe to eat just because natural Bt is believed safe. Studies show that GM Bt crops have toxic effects on non-target insects and mammals.

GM Roundup Ready crops absorb glyphosate herbicide into their tissues with the aid of the adjuvants present in the commercially used formulations. Some of the glyphosate is broken down (metabolized) into a substance called aminomethylphosphonic acid (AMPA). People and animals that eat these GM crops are eating glyphosate and AMPA, both of which are toxic.

Roundup and other glyphosate-based herbicide formulations consist of a mixture of chemicals: the stated "active ingredient" (glyphosate) and an "inert" set of poorly defined "adjuvants". These full commercial herbicide formulations have been shown to be far more toxic than glyphosate alone[1,2] – 125 times more toxic in an in vitro study.[3] Feeding studies in pigs[1] and rats[2] directly comparing the toxicity of formulations with glyphosate alone also found that the formulations were far more toxic.

However, only glyphosate alone is tested for long-term toxicity in safety tests carried out to support regulatory authorizations. The complete formulations are not tested for long-term toxicity for regulatory purposes, even though we are always exposed to the formulations, not just to glyphosate. This is because regulators have accepted the industry position that the adjuvants in pesticide formulations are "inert" and pose no serious health risks.

Following is a list of toxic effects caused by glyphosate, AMPA, and Roundup as revealed in animal studies, laboratory studies in human cells, and human epidemiological and clinical case studies:

- Severe liver and kidney damage[4]
- Chronic kidney disease[5,6]
- Disruption of hormonal systems,[4,7,8,9,10,11,12,13] which can potentially lead to multiple organ damage and hormone-dependent cancer[13]
- Developmental and reproductive toxicity, including damage to sperm[14] and miscarriage and premature birth[15]
- Disruption of beneficial gut bacteria, favouring the growth of botulism-causing bacteria in cows[16]
- Damage to DNA[17,18,19,20,21 22]
- Birth defects[23,24,25]
- Neurotoxicity[26,27,28,29,30]
- Cancer.[31,32,33,34]

It should be remembered that many of the animal studies and laboratory studies in human cells use high, unrealistic doses of glyphosate and Roundup and are of uncertain relevance to human health. However, they demonstrate in-principle toxic effects that need to be validated at environmentally relevant, real-world exposure levels.

In 2015 the World Health Organization's International Agency for Research on Cancer (IARC) declared that glyphosate herbicide is "probably carcinogenic to humans", based largely on compelling data from animal carcinogenicity studies.[35]

Public health crises linked with Roundup exposure

In South America a public health crisis has emerged around the spraying of Roundup herbicide on GM Roundup Ready soy, which is often carried out from the air. The spray drifts into people's homes, schools, food crops, and watercourses. It has been blamed for marked increases in serious health problems, including birth defects and cancer.

A report commissioned by the provincial government of Chaco, Argentina, found that the rate of birth defects increased fourfold and rates of childhood cancers tripled in only a decade in areas where rice and GM soy crops are heavily sprayed with glyphosate and other herbicides.[36] A review of studies on the health effects of pesticides used with GM herbicide-tolerant crops concluded that the precautionary principle was being flouted.[37]

Studies have linked an epidemic of chronic kidney disease in Sri Lanka and other countries to exposure to Roundup. The authors propose that glyphosate becomes toxic to the kidney when it mixes with "hard" water containing metals, or with heavy metals like arsenic and cadmium, where they are naturally present in soil or added as fertilizers. The authors argue that glyphosate binds to these substances and carries them to the kidneys, resulting in tissue destruction.[5,6]

Studies in farm animals may indicate glyphosate accumulation in the body and links to multiple organ damage

Industry and regulators claim that any ingested glyphosate is rapidly cleared mostly via urine and does not accumulate in the body,[38] which supports their position of a high safety profile for this pesticide. However, this view may be contradicted by studies in dairy cows, which show that glyphosate levels in numerous organs, including the liver, kidneys, lung and muscles, are almost as high as those found in urine.[39] This suggests that glyphosate accumulates in the body and/or that constant exposure keeps levels "topped up".

In addition, levels of glyphosate found in cows' urine correlated with blood biochemistry measurements reflective of liver, kidney and muscle damage; that is, the higher the levels of glyphosate found in urine, the greater the indication of damage to these organs.[40]

Glyphosate is everywhere – even in breast milk

Glyphosate turns up everywhere. Glyphosate and its toxic metabolite AMPA were found in over 75% of air and rain samples tested from the Mississippi agricultural region in 2007. The researchers noted that the widespread presence of glyphosate was due to the cultivation of GM glyphosate-tolerant crops.[41]

Glyphosate has been found in women's blood[42] within the range of levels found in vitro to have endocrine disruptive effects on the estrogen hormone system.[13] In an in vitro study simulating human exposures, glyphosate was found to cross the placental barrier and enter the foetal compartment.[43]

Surveys looking at levels of glyphosate and AMPA in the human population are few and limited in scope, but imply that both substances are present at readily detectable levels.[44] Urine levels vary between regions and population groups but are generally much higher in the USA (8- to 10-fold) than in Europe.[44] The first ever survey of human breast milk found levels from 76-166 ppb (parts per billion) in American women. This is 760 to 1600 times higher than the EU permitted level in drinking water. These levels were, however, lower than the 700 μg/L maximum contaminant level (MCL) for glyphosate in drinking water allowed in the US.[45]

The levels of glyphosate found in human urine to date have been claimed by scientists working for the German Federal Institute for Risk Assessment (BfR) not to be a health concern as they indicate a body burden that is well below regulatory set safety limits.[44] However, the amounts detected are within the range found to mimic the action of estrogen in stimulating human breast cancer cell growth.[13] Ignoring potential endocrine disruptive effects of glyphosate as demonstrated in studies such as this represents a major weakness in its safety evaluation.

Glyphosate levels have been found to be significantly higher in the urine of humans who ate non-organic food, compared with those who ate mostly organic food. Chronically ill people showed significantly higher glyphosate residues in their urine than healthy people.[39]

Roundup link with modern diseases suggested

A series of reviews by Samsel and Seneff,[46,47] and Seneff and Swanson,[48] suggested mechanisms by which glyphosate herbicides could be contributing to modern human diseases that are on the increase worldwide, such as celiac disease, gluten intolerance, and Alzheimer's. They cited glyphosate's ability to act as a nutrient metal chelator (binding agent), disrupt gut bacteria, and suppress the activity of detoxifying enzymes, potentially enhancing the damaging effects of other environmental toxins.

These are interesting hypotheses but need to be tested in detailed controlled animal feeding studies using environmentally relevant doses of glyphosate/Roundup to establish whether these chemicals are truly contributing to, or causative in, human disease through these suggested mechanisms.

Are the doses of Roundup that we're exposed to safe?

Many animal tests showing harm from Roundup are conducted with high doses that we would never be exposed to in real life. This leads defenders of pesticides and GM Roundup Ready crops to argue that "the dose makes the poison" and that the doses we are exposed to are below regulatory limits and therefore safe.

However, the regulatory limits vary widely in different jurisdictions and for different crops. The maximum residue limit (MRL) set by regulators for glyphosate in some food and feed crops in the EU is 20 mg/kg.[49] International guidelines set by the UN food safety body Codex allow levels in some animal feed crops up to 500 mg/kg.[50] The "acceptable daily intake" (ADI) of glyphosate is set at 0.3 mg per kg of bodyweight per day (written as 0.3 mg/kg bw/d) within the EU,[51] Australia and New Zealand.[52] The ADI in the US is higher: 1.75 mg/kg bw/d.[53]

Are these levels safe? There is reason for doubt. This is because:

- The supposedly safe levels of glyphosate consumption have never been directly tested to find out if they are safe to consume over the long term. Instead, the supposed safe levels are extrapolated from much higher doses in industry tests that are said not to cause toxic effects. This is not valid because some toxins, especially those that disrupt the hormonal system (endocrine disruptors), are known to exert major toxic effects at low doses as well as at higher doses via different mechanisms. So safe levels cannot be extrapolated from a "no effect" finding at higher doses.[54]

An endocrine disruptive mechanism may be at the basis of the severe organ (liver, kidney, and pituitary) damage, hormonal (testosterone, estrogen) disruption, and trend toward increased mammary tumour incidence in rats fed Roundup at the ultra-low glyphosate equivalent dose of 50ng/L.[4]

- Safety levels are set on the basis of tests on the isolated active ingredient glyphosate, not the complete commercial formulations as sold and used – yet the formulations have consistently been found to be more toxic than the active ingredient alone (see above).
- The data that industry submits to regulators to establish safety limits are not peer-reviewed and published, but are kept secret under commercial confidentiality agreements between industry and regulators. Independent scientists and the public cannot see the details of the research protocols or data in order to evaluate the rigour of the research or the validity of industry's interpretation. Moreover, there is evidence that industry and regulators have abused this situation, downplaying findings of birth defects in laboratory animals in industry studies on glyphosate.[23]

"Extreme" levels of glyphosate found in GM soy

Researchers analyzed the composition of GM glyphosate-tolerant soybeans, industrially grown non-GM soybeans, and organic soybeans, grown in the USA. They found that the GM soybeans contained high residues of glyphosate and its toxic metabolite AMPA (average of 3.3 and 5.7 mg/kg, respectively), but industrially grown non-GM soybeans and organic soybeans contained neither chemical.[55]

Monsanto previously called the levels of glyphosate reported in this study "extreme". It is clear that in GMO agriculture, "extreme" levels of glyphosate have become the norm.[56]

When evaluating these findings, some points need to be considered:

- We do not know how much of the glyphosate in the soy is absorbed by the human or animal consumer when present at these levels in food or feed. So what happened in this experiment on disembodied human cells may not happen in a living human or animal.
- The mean levels found in the soy of 3.3 mg/kg for glyphosate and 5.7 mg/kg for AMPA[55] are below the maximum residue limit set for soy in Europe (20 mg/kg glyphosate). However, that does not mean that these levels are safe to eat, as the residue limit is set for glyphosate alone, not

the complete formulations that we are exposed to, and the formulations are more toxic.

Soybeans grown in Argentina were found in tests to have much higher levels of glyphosate – up to nearly 100 mg/kg. In seven of eleven samples the level was higher than the maximum residue level (MRL) of 20 mg/kg allowed in soybean products used for food and feed.[57]

It can be concluded from these results that people who eat food products from GM Roundup Ready crops are eating amounts of these substances that may have toxic – particularly endocrine disruptive – effects. Further animal testing with low, realistic doses of complete herbicide formulations is needed to confirm or refute this possibility.

GM crops tolerant to 2,4-D and dicamba expand range of health risks

The US Department of Agriculture's approval of GM soybeans and maize tolerant to 2,4-D[58,59] and dicamba[60] as well as glyphosate means that people and animals that eat GM crops will be exposed to the risks posed by all these herbicides and their mixtures. Exposure to 2,4-D has been linked in studies to Non-Hodgkin's lymphoma[61] and soft tissue sarcoma[62] (types of cancer), as well neurological diseases.[63] Dicamba is suspected of causing birth defects.[64] Both herbicides are prone to drift, placing neighbours at risk of exposure. The health effects of mixtures of these herbicides have never been tested for regulatory purposes, as industry only tests the toxicity of a single chemical at a time.

Bt toxin in GM plants is not harmless

Regulators have approved GM Bt crops on the assumption that the GM Bt toxin is the same as natural Bt toxin, a substance derived from a common soil bacterium which is used as an insecticide spray by organic and conventional farmers and said to have a history of safe use. Regulators conclude that GM crops engineered to contain Bt insecticidal protein must also be harmless. However, scientific evidence indicates that this is incorrect.

The Bt toxin expressed by GM Bt plants is different from natural Bt in both structure[65] and mode of action.[66] Due to an unintended outcome during the insertion of the GM gene unit into the plant DNA, Monsanto's GM maize MON810 produces a "truncated" version of the Bt toxin protein – a shorter form of the protein than was intended and thus very different from the natural form.[67] Such changes can result in the protein having unexpected environmental and health effects, since even the change of a single amino acid in a protein can radically change the protein's and therefore the plant's behaviour or properties.[68,69] For instance, the modified protein may be more toxic or allergenic than the natural form.

The differences between natural and GM Bt toxins affect how human and animal consumers might be exposed. Natural Bt is applied to the surface of the leaves and stem of the plant, where it is exposed to sunlight. This rapidly breaks down the Bt toxin, so it is unlikely to be consumed by animals or people that eat the crop. Moreover, natural Bt toxin is produced in an inactive form. It becomes an active toxin only when enzymes in an insect's gut activate it. In contrast, the Bt toxins in GM Bt crops are present in every cell of the GM plant, and the GM Bt toxins are modified so that they are produced in fully active form.[66,70] The preactivated GM Bt toxin lacks the selectivity of natural Bt and is therefore more likely to harm non-target organisms. GM Bt crops and the Bt toxins they are engineered to contain have been found to have toxic effects on butterflies and other non-target insects,[71,72,72] beneficial pest predators,[73,74,75,76,77,78] bees,[79] aquatic organisms,[80,81] and beneficial soil organisms.[82]

Bt toxins and GM Bt crops have toxic effects on mammals

GMO proponents claim that the Bt toxin in GM Bt crops only affects target pests and is harmless to mammals, including people or animals that eat the crops.[83] However, this is false. GM Bt toxins have been found to be toxic to human cells tested in the laboratory.[84] And when GM Bt crops have been fed to mammals in feeding trials, they have had adverse effects, including:

- Toxic effects or signs of toxicity in the small intestine, liver, kidney, spleen, and pancreas[85,86,87,88,89]
- Disturbed functioning of the digestive system[87,89]
- Increased or decreased weight gain compared with controls[85,90]
- Male reproductive organ damage[89]
- Blood biochemistry disturbances[90]
- Immune system disturbances.[91]

GM Bt toxin has been found to produce a potent immune response in the intestines of mice[92,93,94] and to amplify the immune response of mice to other substances.[95]

However, it is important to note that the GM Bt crop feeding studies have not definitively proven that the engineered Bt toxin is the specific cause of harm that has been observed, because the animals have been fed the whole GM Bt crop, which contains thousands of components, not the GM Bt toxin in isolation. What these studies do demonstrate is that GM Bt crops contain substances that cause adverse effects, regardless of the molecular cause of the problem.

Bt toxin found circulating in pregnant women's blood

It is claimed that in mammals, Bt toxins are harmlessly broken down in the digestive tract.[96] This assumption underlies all approvals of GM Bt crops for human and animal feed – yet it is false. A laboratory study simulating human digestion found that the Bt toxin protein was highly resistant to being broken down in realistic stomach acidity conditions and still produced an immune response.[97]

A study conducted in Canada found Bt toxin protein circulating in the blood of pregnant and non-pregnant women and the blood supply to foetuses.[42,98] Whether the Bt toxin originated from GM crops is not known. But wherever it came from, it clearly did not break down fully in the digestive tract.

References

1. Lee H-L, Kan C-D, Tsai C-L, Liou M-J, Guo H-R. Comparative effects of the formulation of glyphosate-surfactant herbicides on hemodynamics in swine. Clin Toxicol Phila Pa. 2009;47(7):651-658. doi:10.1080/15563650903158862.
2. Adam A, Marzuki A, Abdul Rahman H, Abdul Aziz M. The oral and intratracheal toxicities of ROUNDUP and its components to rats. Vet Hum Toxicol. 1997;39(3):147-151.
3. Mesnage R, Defarge N, de Vendomois JS, Séralini GE. Major pesticides are more toxic to human cells than their declared active principles. BioMed Res Int. 2014;2014. doi:10.1155/2014/179691.
4. Séralini G-E, Clair E, Mesnage R, et al. Republished study: long-term toxicity of a Roundup herbicide and a Roundup-tolerant genetically modified maize. Environ Sci Eur. 2014;26(14). doi:10.1186/s12302-014-0014-5.
5. Jayasumana C, Gunatilake S, Senanayake P. Glyphosate, hard water and nephrotoxic metals: Are they the culprits behind the epidemic of chronic kidney disease of unknown etiology in Sri Lanka? Int J Environ Res Public Health. 2014;11(2):2125-2147. doi:10.3390/ijerph110202125.
6. Jayasumana C, Paranagama P, Agampodi S, Wijewardane C, Gunatilake S, Siribaddana S. Drinking well water and occupational exposure to herbicides is associated with chronic kidney disease, in Padavi-Sripura, Sri Lanka. Environ Health. 2015;14(1):6. doi:10.1186/1476-069X-14-6.
7. Séralini G-E, Clair E, Mesnage R, et al. Republished study: long-term toxicity of a Roundup herbicide and a Roundup-tolerant genetically modified maize. Environ Sci Eur. 2014;26(14). doi:10.1186/s12302-014-0014-5.
8. Soso AB, Barcellos LJG, Ranzani-Paiva MJ, et al. Chronic exposure to sub-lethal concentration of a glyphosate-based herbicide alters hormone profiles and affects reproduction of female

Jundiá (Rhamdia quelen). Environ Toxicol Pharmacol. 2007;23:308-313.

9. Walsh LP, McCormick C, Martin C, Stocco DM. Roundup inhibits steroidogenesis by disrupting steroidogenic acute regulatory (StAR) protein expression. Env Health Perspect. 2000;108:769-776.
10. Romano RM, Romano MA, Bernardi MM, Furtado PV, Oliveira CA. Prepubertal exposure to commercial formulation of the herbicide Glyphosate alters testosterone levels and testicular morphology. Arch Toxicol. 2010;84:309-317.
11. Gasnier C, Dumont C, Benachour N, Clair E, Chagnon MC, Séralini GE. Glyphosate-based herbicides are toxic and endocrine disruptors in human cell lines. Toxicology. 2009;262:184-191. doi:10.1016/j.tox.2009.06.006.
12. Hokanson R, Fudge R, Chowdhary R, Busbee D. Alteration of estrogen-regulated gene expression in human cells induced by the agricultural and horticultural herbicide glyphosate. Hum Exp Toxicol. 2007;26:747-752. doi:10.1177/0960327107083453.
13. Thongprakaisang S, Thiantanawat A, Rangkadilok N, Suriyo T, Satayavivad J. Glyphosate induces human breast cancer cells growth via estrogen receptors. Food Chem Toxicol. 2013;59:129-136. doi:10.1016/j.fct.2013.05.057.
14. Dallegrave E, Mantese FD, Oliveira RT, Andrade AJ, Dalsenter PR, Langeloh A. Pre- and postnatal toxicity of the commercial glyphosate formulation in Wistar rats. Arch Toxicol. 2007;81:665-673. doi:10.1007/s00204-006-0170-5.
15. Savitz DA, Arbuckle T, Kaczor D, Curtis KM. Male pesticide exposure and pregnancy outcome. Am J Epidemiol. 1997;146:1025-1036.
16. Krüger M, Shehata AA, Schrödl W, Rodloff A. Glyphosate suppresses the antagonistic effect of Enterococcus spp. on Clostridium botulinum. Anaerobe. 2013;20:74-78.
17. Marc J, Mulner-Lorillon O, Belle R. Glyphosate-based pesticides affect cell cycle regulation. Biol Cell. 2004;96:245-249. doi:10.1016/j.biolcel.2003.11.010.
18. Bellé R, Le Bouffant R, Morales J, Cosson B, Cormier P, Mulner-Lorillon O. Sea urchin embryo, DNA-damaged cell cycle checkpoint and the mechanisms initiating cancer development. J Soc Biol. 2007;201:317-327.
19. Marc J, Mulner-Lorillon O, Boulben S, Hureau D, Durand G, Bellé R. Pesticide Roundup provokes cell division dysfunction at the level of CDK1/cyclin B activation. Chem Res Toxicol. 2002;15(3):326-331.
20. Marc J, Bellé R, Morales J, Cormier P, Mulner-Lorillon O. Formulated glyphosate activates the DNA-response checkpoint of the cell cycle leading to the prevention of G2/M transition. Toxicol Sci. 2004;82:436-442. doi:10.1093/toxsci/kfh281.
21. Mañas F, Peralta L, Raviolo J, et al. Genotoxicity of glyphosate assessed by the Comet assay and cytogenic tests. Env Toxicol Pharmacol. 2009;28:37-41.
22. Mañas F, Peralta L, Raviolo J, et al. Genotoxicity of AMPA, the environmental metabolite of glyphosate, assessed by the Comet assay and cytogenetic tests. Ecotoxicol Env Saf. 2009;72:834-837. doi:10.1016/j.ecoenv.2008.09.019.
23. Antoniou M, Habib MEM, Howard CV, et al. Teratogenic effects of glyphosate-based herbicides: Divergence of regulatory decisions from scientific evidence. J Env Anal Toxicol. 2012;S4:006. doi:10.4172/2161-0525.S4-006.
24. Paganelli A, Gnazzo V, Acosta H, López SL, Carrasco AE. Glyphosate-based herbicides produce teratogenic effects on vertebrates by impairing retinoic acid signaling. Chem Res Toxicol. 2010;23:1586-1595. doi:10.1021/tx1001749.
25. Dallegrave E, Mantese FD, Coelho RS, Pereira JD, Dalsenter PR, Langeloh A. The teratogenic potential of the herbicide glyphosate-Roundup in Wistar rats. Toxicol Lett. 2003;142:45-52.
26. Anadón A, del Pino J, Martínez MA, et al. Neurotoxicological effects of the herbicide glyphosate. Toxicol Lett. 2008;180S:S164.
27. Garry VF, Harkins ME, Erickson LL, Long-Simpson LK, Holland SE, Burroughs BL. Birth defects, season of conception, and sex of children born to pesticide applicators living in the Red River Valley of Minnesota, USA. Env Health Perspect. 2002;110 Suppl 3:441-449.
28. Barbosa ER, Leiros da Costa MD, Bacheschi LA, Scaff M, Leite CC. Parkinsonism after glycine-derivate exposure. Mov Disord. 2001;16:565-568.
29. Wang G, Fan XN, Tan YY, Cheng Q, Chen SD. Parkinsonism after chronic occupational exposure to glyphosate. Park Relat Disord. 2011;17:486-487. doi:10.1016/j.parkreldis.2011.02.003.
30. Gui YX, Fan XN, Wang HM, Wang G, Chen SD. Glyphosate induced cell death through apoptotic and autophagic mechanisms. Neurotoxicol Teratol. 2012;34(3):344-349.
31. George J, Prasad S, Mahmood Z, Shukla Y. Studies on glyphosate-induced carcinogenicity in mouse skin: A proteomic approach. J Proteomics. 2010;73:951-964. doi:10.1016/j.jprot.2009.12.008.
32. Hardell L, Eriksson M. A case-control study of non-Hodgkin lymphoma and exposure to pesticides. Cancer. 1999;85:1353-1360. doi:10.1002/(SICI)1097-0142(19990315)85:6<1353::AID-CNCR19>3.0.CO;2-1.

33. Hardell L, Eriksson M, Nordstrom M. Exposure to pesticides as risk factor for non-Hodgkin's lymphoma and hairy cell leukemia: Pooled analysis of two Swedish case-control studies. Leuk Lymphoma. 2002;43:1043-1049.
34. Eriksson M, Hardell L, Carlberg M, Akerman M. Pesticide exposure as risk factor for non-Hodgkin lymphoma including histopathological subgroup analysis. Int J Cancer. 2008;123:1657-1663. doi:10.1002/ijc.23589.
35. Guyton K, Loomis D, Grosse Y, El Ghissassi F, Benbrahim-Tallaa L. Carcinogenicity of tetrachlorvinphos, parathion, malathion, diazinon, and glyphosate. Lancet Oncol. March 2015. http://www.thelancet.com/pdfs/journals/lanonc/PIIS1470-2045%2815%2970134-8.pdf.
36. Comision Provincial de Investigación de Contaminantes del Agua. Primer Informe [First Report]. Resistencia, Chaco, Argentina; 2010. http://www.gmwatch.org/files/Chaco_Government_Report_Spanish.pdf ; English translation at http://www.gmwatch.org/files/Chaco_Government_Report_English.pdf.
37. Lopez SL, Aiassa D, Benitez-Leite S, et al. Pesticides used in South American GMO-based agriculture: A review of their effects on humans and animal models. In: Fishbein JC, Heilman JM, eds. Advances in Molecular Toxicology. Vol 6. New York: Elsevier; 2012:41-75.
38. Glyphosate Task Force. Residual Traces of Glyphosate in Urine. Glyphosate Task Force; Undated. http://www.glyphosate.eu/system/files/sidebox-files/residual_traces_of_glyphosate_in_urine_en.pdf.
39. Krüger M, Schledorn P, Schrödl W, Hoppe HW, Lutz W, Shehata AA. Detection of glyphosate residues in animals and humans. J Env Anal Toxicol. 2014;4(2). doi:10.4172/2161-0525.1000210.
40. Krüger M, Schrödl W, Neuhaus J, Shehata AA. Field investigations of glyphosate in urine of Danish dairy cows. J Env Anal Toxicol. 2013;3(5). doi:http://dx.doi.org/10.4172/2161-0525.1000186.
41. Majewski MS, Coupe, R. H., Foreman WT, Capel PD. Pesticides in Mississippi air and rain: A comparison between 1995 and 2007. Env Toxicol Chem. February 2014. doi:10.1002/etc.2550.
42. Aris A, Leblanc S. Maternal and fetal exposure to pesticides associated to genetically modified foods in Eastern Townships of Quebec, Canada. Reprod Toxicol. 2011;31(4):528-533.
43. Poulsen MS, Rytting E, Mose T, Knudsen LE. Modeling placental transport: Correlation of in vitro BeWo cell permeability and ex vivo human placental perfusion. Toxicol Vitro. 2009;23:1380-1386. doi:10.1016/j.tiv.2009.07.028.
44. Niemann L, Sieke C, Pfeil R, Solecki R. A critical review of glyphosate findings in human urine samples and comparison with the exposure of operators and consumers. J Für Verbraucherschutz Leb. January 2015:1-10. doi:10.1007/s00003-014-0927-3.
45. Honeycutt Z, Rowlands H. Glyphosate Testing Report: Findings in American Mothers' Breast Milk, Urine and Water.; 2014. http://www.momsacrossamerica.com/glyphosate_testing_results.
46. Samsel A, Seneff S. Glyphosate's suppression of cytochrome P450 enzymes and amino acid biosynthesis by the gut microbiome: Pathways to modern diseases. Entropy. 2013;15:1416-1463.
47. Samsel A, Seneff S. Glyphosate, pathways to modern diseases II: Celiac sprue and gluten intolerance. Interdiscip Toxicol. 2013;6(4):159-184. doi:10.2478/intox-2013-0026.
48. Seneff S, Swanson N, Li C. Aluminum and glyphosate can synergistically induce pineal gland pathology: Connection to gut dysbiosis and neurological disease. Agric Sci. 2015;6:42-70.
49. European Union. EU pesticides database. March 2014. http://ec.europa.eu/sanco_pesticides/public/?event=homepage.
50. Codex Alimentarius. Pesticide Residues in Food and Feed: 158 Glyphosate. Food and Agriculture Organization and World Health Organization; 2013. http://www.codexalimentarius.net/pestres/data/pesticides/details.html?id=158.
51. European Commission Health & Consumer Protection Directorate-General. Review report for the active substance glyphosate. January 2002. http://bit.ly/HQnkFj.
52. Food Standards Australia New Zealand (FSANZ). Assessment of Glyphosate Residues: Supporting Document 2. Canberra, Australia; 2010. http://www.foodstandards.gov.au/code/applications/documents/A1021%20GM%20Maize%20AppR%20SD2%20Glyphosate%20residues%20FINAL.pdf.
53. National Pesticide Information Center. Glyphosate Technical Fact Sheet. Corvallis, OR: NPIC, Oregon State University/US EPA; 2014. http://npic.orst.edu/factsheets/glyphotech.pdf.
54. Vandenberg LN, Colborn T, Hayes TB, et al. Hormones and endocrine-disrupting chemicals: Low-dose effects and nonmonotonic dose responses. Endocr Rev. 2012;33(3):378-455. doi:10.1210/er.2011-1050.
55. Bøhn T, Cuhra M, Traavik T, Sanden M, Fagan J, Primicerio R. Compositional differences in soybeans on the market: glyphosate accumulates in Roundup Ready GM soybeans. Food

Chem. 2013;153(2014):207-215. doi:10.1016/j.foodchem.2013.12.054.
56. Bøhn T, Cuhra M. How "extreme levels" of Roundup in food became the industry norm. Indep Sci News. March 2014. http://www.independentsciencenews.org/news/how-extreme-levels-of-roundup-in-food-became-the-industry-norm/.
57. Testbiotech. High Levels of Residues from Spraying with Glyphosate Found in Soybeans in Argentina. Munich, Germany; 2013. http://www.testbiotech.org/sites/default/files/TBT_Background_Glyphosate_Argentina_0.pdf.
58. Gillam C. Dow launches multi-herbicide tolerant GM soybean. Reuters. http://bit.ly/qBR9a5. Published August 22, 2011.
59. Kimbrell A. "Agent Orange" corn: Biotech only winner in chemical arms race as herbicide resistant crops fail. Huffington Post. http://www.huffingtonpost.com/andrew-kimbrell/agent-orange-corn-biotech_b_1291295.html. Published February 22, 2012.
60. Food & Water Watch. USDA says "yes" to pesticide drift, approves dicamba-tolerant crops. Food & Water Watch. http://www.foodandwaterwatch.org/pressreleases/usda-says-yes-to-pesticide-drift-approves-dicamba-tolerant-crops%E2%80%A8/. Published January 15, 2015.
61. Zahm SH, Weisenburger DD, Babbitt PA, et al. A case-control study of non-Hodgkin's lymphoma and the herbicide 2,4-dichlorophenoxyacetic acid (2,4-D) in eastern Nebraska. Epidemiology. 1990;1:349-356.
62. Hardell L, Eriksson M. The association between soft tissue sarcomas and exposure to phenoxyacetic acids. A new case-referent study. Cancer. 1988;62(3):652-656.
63. Shearer R. Public health effects of the aquatic use of herbicides – 2,4-D, dichlobenil, endothall and diquat. In: Shearer R, Halter M, eds. Literature Reviews of Four Selected Herbicides: 2,4-D, Dichlobenil, Diquat & Endothall. Vol Seattle, WA: Municipality of Metropolitan Seattle; 1980.
64. PubChem Open Chemistry Database. Dicamba. PubChem Open Chem Database. 2015. http://pubchem.ncbi.nlm.nih.gov/compound/dicamba#section=Top.
65. Séralini GE, Mesnage R, Clair E, Gress S, de Vendômois JS, Cellier D. Genetically modified crops safety assessments: Present limits and possible improvements. Environ Sci Eur. 2011;23. doi:10.1186/2190-4715-23-10.
66. Székács A, Darvas B. Comparative aspects of Cry toxin usage in insect control. In: Ishaaya I, Palli SR, Horowitz AR, eds. Advanced Technologies for Managing Insect Pests. Vol Dordrecht, Netherlands: Springer; 2012:195-230.
67. Freese W, Schubert D. Safety testing and regulation of genetically engineered foods. Biotechnol Genet Eng Rev. 2004:299-324.
68. Doyle MR, Amasino RM. A single amino acid change in the enhancer of zeste ortholog CURLY LEAF results in vernalization-independent, rapid flowering in Arabidopsis. Plant Physiol. 2009;151:1688-1697. doi:10.1104/pp.109.145581.
69. Hanzawa Y, Money T, Bradley D. A single amino acid converts a repressor to an activator of flowering. Proc Natl Acad Sci U A. 2005;102:7748-7753. doi:10.1073/pnas.0500932102.
70. Li H, Buschman LL, Huang F, Zhu KY, Bonning B, Oppert BA. Resistance to Bacillus thuringiensis endotoxins in the European corn borer. Biopestic Int. 2007;3:96-107.
71. Losey JE, Rayor LS, Carter ME. Transgenic pollen harms monarch larvae. Nature. 1999;399:214. doi:10.1038/20338.
72. Jesse LCH, Obrycki JJ. Field deposition of Bt transgenic corn pollen: Lethal effects on the monarch butterfly. J Oecologia. 2000;125:241-248.
73. Hilbeck A, Baumgartner M, Fried PM, Bigler F. Effects of transgenic Bt corn-fed prey on immature development of Chrysoperla carnea (Neuroptera: Chrysopidae). Environ Entomol. 1998;27(2):480-487.
74. Hilbeck A, McMillan JM, Meier M, Humbel A, Schlaepfer-Miller J, Trtikova M. A controversy re-visited: Is the coccinellid Adalia bipunctata adversely affected by Bt toxins? Environ Sci Eur. 2012;24(10). doi:10.1186/2190-4715-24-10.
75. Hilbeck A, Meier M, Trtikova M. Underlying reasons of the controversy over adverse effects of Bt toxins on lady beetle and lacewing larvae. Environ Sci Eur. 2012;24(9). doi:10.1186/2190-4715-24-9.
76. Hilbeck A, Moar WJ, Pusztai-Carey M, Filippini A, Bigler F. Prey-mediated effects of Cry1Ab toxin and protoxin and Cry2A protoxin on the predator Chrysoperla carnea. Entomol Exp Appl. 1999;91:305-316.
77. Marvier M, McCreedy C, Regetz J, Kareiva P. A meta-analysis of effects of Bt cotton and maize on nontarget invertebrates. Science. 2007;316:1475-1477. doi:10.1126/science.1139208.
78. Lövei GL, Arpaia S. The impact of transgenic plants on natural enemies: A critical review of laboratory studies. Entomol Exp Appl. 2005;114:1-14. doi:10.1111/j.0013-8703.2005.00235.x.
79. Ramirez-Romero R, Desneux N, Decourtye A, Chaffiol A, Pham-Delègue MH. Does Cry1Ab protein affect learning performances of the honey bee Apis mellifera L. (Hymenoptera, Apidae)? Ecotoxicol Environ Saf. 2008;70:327-333.

80. Rosi-Marshall EJ, Tank JL, Royer TV, et al. Toxins in transgenic crop byproducts may affect headwater stream ecosystems. Proc Natl Acad Sci USA. 2007;104:16204-16208. doi:10.1073/pnas.0707177104.
81. Bøhn T, Traavik T, Primicerio R. Demographic responses of Daphnia magna fed transgenic Bt-maize. Ecotoxicology. 2010;19:419-430. doi:10.1007/s10646-009-0427-x.
82. Castaldini M, Turrini A, Sbrana C, et al. Impact of Bt corn on rhizospheric and soil eubacterial communities and on beneficial mycorrhizal symbiosis in experimental microcosms. Appl Env Microbiol. 2005;71:6719-6729. doi:10.1128/AEM.71.11.6719-6729.2005.
83. GMO Compass. Environmental safety: insects, spiders, and other animals. December 2006. http://www.gmo-compass.org/eng/safety/environmental_safety/169.effects_gm_plants_insects_spiders_animals.html.
84. Mesnage R, Clair E, Gress S, Then C, Székács A, Séralini G-E. Cytotoxicity on human cells of Cry1Ab and Cry1Ac Bt insecticidal toxins alone or with a glyphosate-based herbicide. J Appl Toxicol. May 2011. http://www.ncbi.nlm.nih.gov/pubmed/22337346.
85. Séralini GE, Cellier D, Spiroux de Vendomois J. New analysis of a rat feeding study with a genetically modified maize reveals signs of hepatorenal toxicity. Arch Environ Contam Toxicol. 2007;52:596-602.
86. De Vendomois JS, Roullier F, Cellier D, Séralini GE. A comparison of the effects of three GM corn varieties on mammalian health. Int J Biol Sci. 2009;5:706-726.
87. Trabalza-Marinucci M, Brandi G, Rondini C, et al. A three-year longitudinal study on the effects of a diet containing genetically modified Bt176 maize on the health status and performance of sheep. Livest Sci. 2008;113:178-190. doi:10.1016/j.livsci.2007.03.009.
88. Fares NH, El-Sayed AK. Fine structural changes in the ileum of mice fed on delta-endotoxin-treated potatoes and transgenic potatoes. Nat Toxins. 1998;6(6):219-233.
89. El-Shamei ZS, Gab-Alla AA, Shatta AA, Moussa EA, Rayan AM. Histopathological changes in some organs of male rats fed on genetically modified corn (Ajeeb YG). J Am Sci. 2012;8(10):684-696.
90. Gab-Alla AA, El-Shamei ZS, Shatta AA, Moussa EA, Rayan AM. Morphological and biochemical changes in male rats fed on genetically modified corn (Ajeeb YG). J Am Sci. 2012;8(9):1117-1123.
91. Finamore A, Roselli M, Britti S, et al. Intestinal and peripheral immune response to MON810 maize ingestion in weaning and old mice. J Agric Food Chem. 2008;56:11533-11539. doi:10.1021/jf802059w.
92. Vázquez-Padrón RI, Moreno-Fierros L, Neri-Bazan L, de la Riva GA, Lopez-Revilla R. Intragastric and intraperitoneal administration of Cry1Ac protoxin from Bacillus thuringiensis induces systemic and mucosal antibody responses in mice. Life Sci. 1999;64:1897-1912.
93. Vázquez-Padrón RI, Moreno-Fierros L, Neri-Bazan L, Martinez-Gil AF, de-la-Riva GA, Lopez-Revilla R. Characterization of the mucosal and systemic immune response induced by Cry1Ac protein from Bacillus thuringiensis HD 73 in mice. Braz J Med Biol Res. 2000;33:147-155.
94. Vázquez-Padrón RI, Gonzales-Cabrera J, Garcia-Tovar C, et al. Cry1Ac protoxin from Bacillus thuringiensis sp. kurstaki HD73 binds to surface proteins in the mouse small intestine. Biochem Biophys Res Commun. 2000;271:54-58. doi:10.1006/bbrc.2000.2584.
95. Vázquez-Padron RI, Moreno-Fierros L, Neri-Bazan L, De La Riva GA, Lopez-Revilla R. Bacillus thuringiensis Cry1Ac protoxin is a potent systemic and mucosal adjuvant. Scand J Immunol. 1999;49:578-584.
96. US Environmental Protection Agency (EPA). Bt Plant-Incorporated Protectants: October 15, 2001 Biopesticides Registration Action Document. Washington, DC: US Environmental Protection Agency (EPA); 2001. http://www.epa.gov/oppbppd1/biopesticides/pips/bt_brad2/7-coorn.pdf.
97. Guimaraes V, Drumare MF, Lereclus D, et al. In vitro digestion of Cry1Ab proteins and analysis of the impact on their immunoreactivity. J Agric Food Chem. 2010;58:3222-3231. doi:10.1021/jf903189j.
98. Aris A. Response to comments from Monsanto scientists on our study showing detection of glyphosate and Cry1Ab in blood of women with and without pregnancy. Reprod Toxicol. 2012;33:122-123.

11 Myth: GM herbicide-tolerant crops are environmentally friendly

Truth: GM herbicide-tolerant crops are an extension of chemical-intensive agriculture and pose threats to the environment

Myth at a glance

GMO proponents make inflated claims for the environmental friendliness of GM herbicide-tolerant crops, but these do not stand up to scrutiny.

For example, it is claimed that GM Roundup Ready (RR) crops are climate-friendly because they allow farmers to adopt the no-till cultivation system of cultivation, in which weeds are controlled with herbicides rather than ploughing. No-till conserves soil and water and is claimed to reduce emissions of the climate change gas carbon dioxide by sequestering more carbon in the soil.

However, the data show that

- ➔ The introduction of GM crops in the US did not significantly increase no-till adoption
- ➔ Once the ecological damage caused by herbicides is taken into account, GM soy is worse for the environment than non-GM soy in no-till and tillage systems
- ➔ No-till fields do not sequester more carbon than ploughed fields when soil depths greater than 30 cm are taken into account.

The UK government's farm-scale trials showed that GM herbicide-tolerant crops are generally worse for biodiversity than non-GM crops grown under conventional chemical-intensive management.

GMO proponents claim that GM herbicide-tolerant crops, notably GM Roundup Ready (RR) crops, are environmentally friendly. Proponents say that GM RR crops allow farmers to adopt the no-till system of cultivation, in which weeds are controlled through herbicide applications rather than by ploughing. No-till conserves soil and water and is claimed to reduce emissions of the climate change gas carbon dioxide by storing (sequestering) more carbon in the soil.

However, there are several problems with this argument:

- GM is not needed for no-till: Farmers do not have to adopt GM crops or use herbicides to practise no-till, which is also used in chemically-based non-GM and agroecological farming.
- Adoption of GM crops in the US did not significantly increase the adoption of no-till, according to the US Department of Agriculture. Adoption of no-till and low-till for soybeans grew from 25% of the soybean acreage in 1990 to 48% in 1995, the 5-year period previous to the introduction of GM herbicide-tolerant soybeans. Growth of no-till and low-till increased further in 1996, the year herbicide-tolerant soybeans were introduced, but then stagnated to 50–60% of all soybean acres in the following years.[1]
- No-till with GM crops is not environmentally friendly because of the herbicides used. A study comparing the environmental impacts of growing GM RR and non-GM soy in Argentina found the negative environmental impact of GM soy was higher than that of non-GM soy in both no-till and tillage systems, because of the ecological impact of the herbicides used. The adoption of no-till was worse for the environment, whether the soy was GM or non-GM. The main reason for the increase in herbicides in no-till systems was the spread of glyphosate-resistant superweeds.[2] In 2015 an article in the US farm press reported that hundreds of thousands of no-till or low-till acres might have to be converted to higher-intensity tillage systems in an effort to control superweeds. USDA weed scientist Andrew Price recommended farmers to plant weed-smothering cover crops, a traditional agroecological method of weed control.[3]
- No-till fields sequester no more carbon than ploughed fields when carbon sequestration at soil depths greater than 30 cm is taken into account, according to a comprehensive review of the scientific literature.[4]

GM herbicide-tolerant crops worse for biodiversity than non-GM chemically grown crops

In the early 2000s the UK government carried out farm-scale trials to test the effects on biodiversity of GM crop management compared with conventional chemically intensive non-GM crop management. The following GM crops were grown:

- Roundup Ready sugar beet and fodder beet
- Glufosinate ammonium-tolerant winter oilseed rape (canola)
- Glufosinate ammonium-tolerant spring oilseed rape
- Glufosinate ammonium-tolerant fodder maize.[5]

The researchers investigated whether the changes in weed management associated with herbicide-tolerant GM crops would reduce weed levels and have wider impacts on farmland biodiversity.[5] The direct toxic effects of herbicides on wildlife were not studied.

The results for beet and oilseed rape showed that GM herbicide-tolerant crop management significantly reduced weeds and weed seeds and therefore would further damage farmland wildlife.[5,6,7,8,9,10,11,12,13] For maize the results showed GM herbicide-tolerant crop management to be better for wildlife than conventional chemically intensive management.[5,6,7,8,9,10,11,12,13] However, the conventional weed control used the highly toxic herbicide atrazine, which was banned in Europe before the results of the farm-scale trials were published.[14]

A more useful comparator for the GM herbicide-tolerant maize would have been maize grown in an integrated pest management (IPM) system, which could reduce or eliminate herbicide use.

The outcome of the farm-scale trials was that no GM crops were approved for UK cultivation.[14] At the time of writing, the UK government appears to have forgotten the results of its own research and is pushing for GM crops to be grown in England,[15] even though the only GM crops in the EU approvals pipeline that could be grown in England are herbicide-tolerant.

References

1. Fernandez-Cornejo J, McBride WD. The Adoption of Bioengineered Crops. Agricultural Economic Report No. 810. Washington, DC: US Department of Agriculture; 2002.
2. Bindraban PS, Franke AC, Ferrar DO, et al. GM-Related Sustainability: Agro-Ecological Impacts, Risks and Opportunities of Soy Production in Argentina and Brazil. Wageningen, the Netherlands: Plant Research International; 2009. http://bit.ly/Ink59c.
3. Hollis P. Conservation tillage systems threatened by herbicide-resistant weeds. Southeast Farm Press. http://southeastfarmpress.com/management/conservation-tillage-systems-threatened-herbicide-resistant-weeds?page=1. Published March 11, 2015.
4. Baker JM, Ochsner TE, Venterea RT, Griffis TJ. Tillage and soil carbon sequestration – What do we really know? Agric Ecosyst Environ. 2007;118:1-5.
5. DEFRA. Managing GM Crops with Herbicides: Effects on Farmland Wildlife. Farmscale Evaluations Research Consortium and the Scientific Steering Committee; 2005. http://bit.ly/P8ocOW.
6. Hawes C, Haughton AJ, Osborne JL, et al. Responses of plants and invertebrate trophic groups to contrasting herbicide regimes in the Farm Scale Evaluations of genetically modified herbicide-tolerant crops. Philos Trans R Soc Lond B Biol Sci. 2003;358:1899-1913. doi:10.1098/rstb.2003.1406.
7. Roy DB, Bohan DA, Haughton AJ, et al. Invertebrates and vegetation of field margins adjacent to crops subject to contrasting herbicide regimes in the Farm Scale Evaluations of genetically modified herbicide-tolerant crops. Philos Trans R Soc Lond B Biol Sci. 2003;358:1879-1898. doi:10.1098/rstb.2003.1404.
8. Brooks DR, Bohan DA, Champion GT, et al. Invertebrate responses to the management of genetically modified herbicide-tolerant and conventional spring crops. I. Soil-surface-active invertebrates. Philos Trans R Soc Lond B Biol Sci. 2003;358:1847-1862. doi:10.1098/rstb.2003.1407.
9. Heard MS, Hawes C, Champion GT, et al. Weeds in fields with contrasting conventional and genetically modified herbicide-tolerant crops. II. Effects on individual species. Philos Trans R Soc Lond B Biol Sci. 2003;358:1833-1846. doi:10.1098/rstb.2003.1401.
10. Firbank LG. Introduction: The farm scale evaluations of spring-sown genetically modified crops. Phil Trans R Soc Lond. 2003;358:1777-1778.
11. Bohan DA, Boffey CW, Brooks DR, et al. Effects on weed and invertebrate abundance and diversity of herbicide management in genetically modified herbicide-tolerant winter-sown oilseed rape. Proc Biol Sci. 2005;272:463-474. doi:10.1098/rspb.2004.3049.
12. BBC News. Q&A: GM farm-scale trials. http://news.bbc.co.uk/2/hi/science/nature/3194574.stm. Published March 9, 2004.
13. Amos J. GM study shows potential "harm." BBC News. http://news.bbc.co.uk/1/hi/sci/tech/4368495.stm. Published March 21, 2005.
14. Friends of the Earth. Press Briefing: Government to Publish the Final Results of the Farm Scale Evaluations of Genetically Modified Crops: Winter Oilseed Rape. London, UK; 2004. http://www.foe.co.uk/sites/default/files/downloads/government_to_publish_the.pdf.
15. Neslen A. GM crops to be fast-tracked in UK following EU vote. The Guardian. http://www.theguardian.com/environment/2015/jan/13/gm-crops-to-be-fast-tracked-in-uk-following-eu-vote. Published January 13, 2015.

12 Myth: GM crops can "coexist" with non-GM and organic crops

Truth: Coexistence means widespread contamination of non-GM and organic crops

Myth at a glance

"Coexistence" of GM with non-GM and organic crops inevitably results in GM contamination of the non-GM and organic crops. This removes choice from farmers and consumers, forcing everyone to produce and consume crops that are potentially GM-contaminated into the indefinite future.

GM contamination cannot be recalled. On the contrary, as GMOs are living organisms, they are likely to persist and proliferate.

There have been numerous GM contamination events since GMOs were first released, since the GMO industry cannot control the spread of its patented GM genes. These contamination events have cost the food and GMO industry and the US government millions of dollars in lost markets, legal damages and compensation schemes for producers, and product recalls.

GM crops do not have to be grown commercially to cause contamination. There have been numerous incidents of GM contamination of non-GM crops and foods from supposedly controlled field trials and restricted releases.

Many of these contamination incidents occurred in the US, demonstrating that even a technologically advanced nation cannot protect its agricultural system, food supply, and export markets from GM contamination.

The GMO lobby argues for "coexistence" of GM, non-GM conventionally farmed, and organic crops. They say that farmers should be able to choose to plant GM crops if they wish and imply that no serious problems would be caused for non-GM and organic farmers.[1]

But experience has shown that the arrival of GM crops in a country removes choice. "Coexistence" rapidly results in widespread contamination of non-GM crops, resulting in lost markets. Contamination occurs through cross-pollination, spread of GM seed by farm machinery, and inadvertent mixing during storage. Farmers are gradually forced to grow GM crops or have their non-GM crops contaminated.

Scientific surveys confirm that GM contamination is unavoidable once GM crops are grown in a region. For example, GM herbicide-tolerant oilseed rape (canola) "volunteers" – plants that were not deliberately planted but are the result of the shedding of seeds from GM crops previously grown in the field – were found growing ten years after the GM oilseed rape crop had been planted.[2] GM herbicide-resistant oilseed rape was found thriving in the wild in North Dakota, often far from areas of agricultural production. GM genes were present in 80% of the wild canola plants found.[3,4]

As no official body keeps records of GM contamination incidents, the NGOs Greenpeace and GeneWatch UK have stepped into the gap with their GM Contamination Register.[5] In the years 2005–2007 alone, 216 contamination incidents were recorded in the database.[6]

Who is liable for GM contamination?

In most countries, legal liability for GM contamination is not clearly established and farmers have had to fight their case in the courts. But in Germany, a law has been passed making farmers who grow GM crops liable for economic damages to non-GM and organic farmers resulting from GM contamination.[7,8] The planting of GM crops in the country rapidly declined and had been abandoned by 2012.[9] The fact that farmers who previously chose to grow GM crops have ceased to do so because they could be held liable for damages implies that coexistence is impossible.

GMO contamination incidents from field trials and supposedly restricted releases

GM crops do not have to be grown commercially to cause contamination. There have been numerous incidents of GM contamination of non-GM crops and foods from supposedly controlled field trials and restricted releases. The key conclusions from collective experience with field trials are:

- Researchers are unable to control the spread of experimental GMOs beyond the test plot, making contamination from field trials to nearby commercial crops nearly inevitable.

- Contamination of commercial seed and crops from even small field trials has led over and over again to contamination incidents that have resulted in multi-billion dollar losses to the agricultural and food industries.

GM contamination: The learning process

"OK, we know that cross-pollination will occur but we've got thirty years of experience to say we know how far pollen will travel. And therefore what we've done is we'll grow a GM crop at a distance away from a non-GM crop, so the people that want non-GM can buy non-GM, and the people that want GM can buy GM. The two will not get mixed up. Everybody will have the right to choose."

– Paul Rylott, seed manager for Aventis CropScience (now Bayer), BBC television broadcast, 2000[10]

"If some people are allowed to choose to grow, sell and consume GM foods, soon nobody will be able to choose food, or a biosphere, free of GM. It's a one way choice, like the introduction of rabbits or cane toads to Australia; once it's made, it can't be reversed."

– Roger Levett, specialist in sustainable development, 2008[11]

"There are... clear data that some consignments of identity-preserved and organic commodities have been tested and found to contain GE material in amounts that exceed contractual requirements or de facto market standards. Such rejected shipments pose problems for those farmers whose loads have been rejected."

– USDA Advisory Committee on Biotechnology and 21st Century Agriculture, 2012[12]

US rice contaminated: $1 billion price tag

In 2006 a GM rice variety, LL601, which was only grown in field trials in the US and was not at the time approved for commercialization, was found to have contaminated the US rice supply and seed stocks.[13] Contaminated rice was found as far afield as Africa, Europe, and Central America. In order to calm market fears, the US Department of Agriculture deregulated the rice in late 2006.[14,15] However, no regulatory authority anywhere else in the world approved it for cultivation or import. The European Food Safety Authority (EFSA) said there was not enough safety data to complete a risk assessment.[16] In 2007 US rice exports dropped 20% as a result of the contamination.[17] In

2011 Bayer, which had acquired the developer company Aventis, agreed to pay $750 million in damages to 11,000 US farmers whose rice crops were contaminated.[18] A court ordered Bayer to pay $137 million in damages to Riceland, a rice export company, for loss of sales to the EU.[19] The worldwide estimated total economic loss due to the LL601 contamination event, over and above the Bayer damages, is estimated at up to $1.3 billion.[13]

China's rice exports impacted by experimental GM rice

In 2011 a GM Bt insecticidal rice, Bt63, was found in baby formula and rice noodles on sale in China.[20] Bt63 was only grown in experimental field trials. It has not been shown to be safe to eat. Contaminated rice products were found in Germany,[21] Sweden,[22] and New Zealand, where the discovery led to product recalls.[23] In 2008 the European Union (EU) implemented emergency measures against Chinese rice imports, requiring compulsory certification that they do not contain Bt63.[24] In 2011 four illegal GM rice traits were found in Chinese consignments at EU ports, and further controls were put in place on Chinese imports.[25] Bt63 contamination of rice imports into the EU was still being reported in 2012.[26]

India and Pakistan Basmati rice rejected by EU due to GM contamination

In 2012 the EU rejected a number of Basmati rice shipments from India and Pakistan because they were contaminated with unauthorized GM varieties.[27]

GM flax causes collapse of Canadian export market

In the 1990s Canada and the US approved a GM flax variety called CDC Triffid. It was never commercialized and was withdrawn in 2001 by the Canadian authorities after flax growers complained that their exports to Europe would be damaged by GM contamination.[28]

Nevertheless, in 2009 Triffid was found in Canadian flax seed supplies, resulting in the EU halting flax imports from Canada.[28,29,30] Imports later resumed, but did not recover to the levels prior to the GM contamination. In 2011 around 4% of samples tested positive for Triffid. Will Hill, president of the Flax Council of Canada, said there was "very little, if any" Canadian flaxseed moving into Europe as the risk of contamination was too high. In 2011–12 Canadian farmers grew 378,500 tonnes of flaxseed, only 41% of the amount grown in the year prior to the Triffid issue.[31]

Possible GMO allergen found in food supply 10 years after being withdrawn

A GM Bt maize called StarLink was found to have contaminated the US food supply in 2000, shutting export markets, forcing the recall of hundreds of food products, and costing approximately $1 billion.[32]

This was not an example of a field trial contamination case, but of a partial release that had taken place under the false assumption that GM genes could be controlled and contained. Regulators had allowed StarLink to be grown for animal feed and industrial use but had not approved it for human food because of suspicions that the Bt insecticidal protein it contained might cause allergic reactions. The US Centers for Disease Control (CDC) carried out tests on blood serum taken from a small number of those people who had reported reactions and concluded that there was no evidence that StarLink was the cause.[33] However, an expert panel convened by the US Environmental Protection Agency (EPA) challenged the methodology and sensitivity of CDC's study and concluded that there was a "medium likelihood" that the Bt protein was an allergen.[34]

The company that developed StarLink, Aventis, withdrew the variety in 2000.[35] But in an example of the impossibility of recalling a GMO once it has been released, it was still detected in samples gathered from Saudi Arabian markets in 2009 and 2010.[36]

Hawaiian papaya contaminated by GM field trials

GM papaya was commercialized in Hawaii in 1998. In 2004 massive unintentional GM contamination of papaya seeds of around 50% was found in Hawaii Island (Big Island). Fruit from an organic farm was contaminated at around 5%. The University of Hawaii's non-GMO seed stocks were found to be contaminated. One source of contamination was the poorly controlled GM papaya field trial site, run by the University of Hawaii. The site was in the middle of the papaya growing area. Spread of GM genes was assisted by theft of fruit because the site was not guarded.[37]

The group that arranged independent laboratory testing of the samples, GMO Free Hawaii (renamed Hawaii SEED in 2006), concluded, "Even as a field trial, GMO papaya cannot be contained." The group added, "Most concerning has been the loss of lucrative export and organic markets caused by the GMO papaya contamination".[37]

Kona coffee producers oppose GM field trials

Aware that field trials can cause severe economic harm, the producers of

Kona coffee in 2008 successfully lobbied for a ban on any cultivation of GM coffee, including in field trials, on the grounds that GM contamination would result in a loss of value for the product.[38,39]

GM wheat grown only in field trials hits US export markets

In 2013 GM Roundup Ready wheat was found growing in a field in Oregon, USA.[40] In response to the discovery, Japan and South Korea suspended US wheat imports for months.[43] GM wheat has never been commercialized anywhere in the world, due to strong market rejection. The last approved field trial of Roundup Ready wheat in Oregon was in 2001.[42] In 2014 the US Department of Agriculture (USDA) closed its investigation into the contamination, saying it had been unable to discover the source. At the same time it opened a new investigation into a second incident in which GM wheat was found growing in Montana in 2014, at a location where Monsanto last trialled the wheat in 2003.[43]

Organic markets damaged by GMO crop cultivation

GM contamination of crops has impacted the livelihoods of farmers who grow organic crops:

- In Canada, contamination from GM oilseed rape has made it virtually impossible to cultivate organic non-GM oilseed rape.[44]
- Organic maize production in Spain has dropped as the acreage of GM maize production has increased, due to contamination by cross-pollination with GM maize.[45]

USDA inspector criticizes oversight of GM field trials

A report by the USDA Office of Inspector General in 2005 harshly criticized the oversight of GM field trials by the USDA's Animal and Plant Health Inspection Service (APHIS) over "weaknesses" in controls which could lead to GMOs persisting in the environment before they are deemed safe to grow.[46]

Nine years on, the situation had not improved. An investigation in the US by Hearst Newspapers found numerous examples of mistakes and violations in GM field trials:[47,48]

- Minimal penalties: The USDA issued just two civil penalties for field trials infractions since 2010 despite sending out nearly 200 notices of non-compliance – incidents from minor paperwork violations to lost seeds to GM plants sprouting where they should not.[47]
- Monsanto mistakes: Monsanto received at least 35 notices of non-compliance with field trial rules from 2010 to 2013. In 2010 the company paid a civil penalty after accidentally ginning experimental cotton in Texas, an error that led to unapproved cottonseed meal and hulls being consumed by livestock and exported to Mexico for animal feed.[47]
- California "pharming": GM pharmaceutical-producing maize is being grown by a company whose founder and president previously founded another company that was permanently banned from running GM trials after creating serious contamination events. These events led to the need to destroy more than half a million bushels of soybeans and more than 150 acres of corn.[48]
- Inadequate buffer distances: Experimental GM apple trees in Washington were allowed to flower less than 100 feet from non-GM apple trees.[47]
- Researcher negligence: The University of Florida received a letter of complaint from the USDA's APHIS after a GM tomato researcher told an inspector he didn't plan to monitor adjacent land for unwanted volunteer plants and intended to lie about it if asked.[47]
- Bad weather: Heavy rains washed out or otherwise damaged test plots, raising the spectre of unwanted dispersal of GMOs.[47]
- Creatures: Among more than 30 incidents, pigs in Texas destroyed a plot of GM sugarcane; wild pigs in Hawaii preyed on Monsanto GM maize; and in Iowa, cows ate still unapproved GM maize plants after entering a gate that was "inadvertently left open".[47]

References

1. SCIMAC (Supply Chain Initiative on Modified Agricultural Crops. GM crop co-existence in perspective. 2006. http://www.scimac.org.uk/files/GM_crop_%20coexistence_perspective.pdf.
2. D'Hertefeldt T, Jørgensen RB, Pettersson LB. Long-term persistence of GM oilseed rape in the seedbank. Biol Lett. 2008;4:314-317.
3. Gilbert N. GM crop escapes into the American wild. Nature. August 2010. http://www.nature.com/news/2010/100806/full/news.2010.393.html.
4. Black R. GM plants "established in the wild." BBC News. http://www.bbc.co.uk/news/science-environment-10859264. Published August 6, 2010.
5. Greenpeace and GeneWatch UK. GM contamination register. 2014. http://www.gmcontaminationregister.org/index.php?content=ho.
6. Greenpeace and GeneWatch UK. GM Contamination Register Report 2007. Amsterdam, The

Netherlands: Greenpeace International; 2008. http://www.greenpeace.org/international/Global/international/planet-2/report/2008/2/gm-contamination-register-2007.pdf.
7. Bhattacharya S. German farmers to be liable for GM contamination. New Sci. November 2004. http://www.newscientist.com/article/dn6729-german-farmers-to-be-liable-for-gm-contamination.html.
8. Hogan M, Niedernhoefer D. German court upholds GMO planting curbs. Reuters. http://uk.reuters.com/article/2010/11/24/us-germany-gmo-idUSTRE6AN55420101124. Published November 24, 2010.
9. Friends of the Earth Europe. GM crops irrelevant in Europe. February 2013. http://www.stopthecrop.org/sites/default/files/content/attachments/foee_factsheet_feb_2013_gmcrops_irrelevant_in_europe.pdf.
10. Rylott P. Matter of Fact [television broadcast]. BBC2 Eastern Region. October 12, 2000.
11. Levett R. Choice: Less can be more. Food Ethics. 2008;3(3):11.
12. USDA Advisory Committee on Biotechnology and 21st Century Agriculture (AC21). Enhancing Coexistence: A Report of the AC21 to the Secretary of Agriculture. Washington, DC; 2012. http://www.usda.gov/documents/ac21_report-enhancing-coexistence.pdf.
13. Blue EN. Risky Business: Economic and Regulatory Impacts from the Unintended Release of Genetically Engineered Rice Varieties into the Rice Merchandising System of the US. Greenpeace; 2007. http://www.greenpeace.org/australia/PageFiles/351482/risky-business.pdf.
14. US Department of Agriculture (USDA). USDA deregulates lines of genetically engineered rice. November 2006. http://www.aphis.usda.gov/newsroom/content/2006/11/rice_deregulate.shtml.
15. Delta Farm Press. Deregulation of LL601 was aboveboard, says USDA. http://deltafarmpress.com/deregulation-ll601-was-aboveboard-says-usda. Published January 22, 2007.
16. European Food Safety Authority (EFSA). EFSA Press Release: EFSA's GMO Panel provides reply to European Commission request on GM rice LLRICE601. September 2006. http://www.efsa.europa.eu/en/press/news/gmo060915.htm.
17. Reuters. Mexico halts US rice over GMO certification. http://www.gmwatch.org/latest-listing/1-news-items/3625. Published March 16, 2007.
18. Harris A, Beasley D. Bayer agrees to pay $750 million to end lawsuits over gene-modified rice. Bloomberg. http://www.bloomberg.com/news/2011-07-01/bayer-to-pay-750-million-to-end-lawsuits-over-genetically-modified-rice.html. Published July 2, 2011.
19. Fox JL. Bayer's GM rice defeat. Nat Biotechnol. 2011;29(473). http://www.nature.com/nbt/journal/v29/n6/full/nbt0611-473c.html.
20. Greenpeace. Children and infants in China at risk of eating food contaminated by illegal GE rice. http://www.greenpeace.org/eastasia/press/releases/food-agriculture/2011/ge-rice-baby-food/. Published April 20, 2011.
21. Greenpeace and GeneWatch UK. Germany finds unauthorised genetically modified (Bt63) rice noodles. GM Contamination Register. http://bit.ly/1nEKmEO. Published June 15, 2011.
22. Greenpeace and GeneWatch UK. Sweden finds unauthorised genetically modified (Bt63) rice. GM Contamination Register. http://bit.ly/1kXDCSP. Published June 27, 2011.
23. New Zealand Food Safety Authority (NZFSA). Unauthorised GM rice product found and withdrawn. http://www.foodsafety.govt.nz/elibrary/industry/Unauthorised_Rice-Zealand_Food.htm. Published July 30, 2008.
24. European Commission. Commission requires certification for Chinese rice products to stop unauthorised GMO from entering the EU. February 2008. http://europa.eu/rapid/press-release_IP-08-219_en.htm.
25. European Commission. Final Report of an Audit Carried out in China from 29 March to 8 April 2011 in Order to Evaluate the Control Systems for Genetically Modified Organisms (GMOs) in Respect of Seed, Food and Feed Intended for Export to the EU. Brussels, Belgium; 2011.
26. Eurofins. New regulations concerning GMO rice from China. Eurofins Food Testing Newsletter No. 38. http://www.eurofins.de/food-analysis/information/food-testing-newsletter/food-newsletter-38/gmo-rice-from-china.aspx. Published March 2012.
27. Colombini D. Food manufacturers warned of GMO rice fraud. FoodManufacture.co.uk. April 2012. http://www.foodmanufacture.co.uk/Food-Safety/Food-manufacturers-warned-of-GMO-rice-fraud.
28. Canadian Grain Commission. Background information on genetically modified material found in Canadian flaxseed. February 2010. http://www.grainscanada.gc.ca/gmflax-lingm/pfsb-plcc-eng.htm.
29. Dawson A. CDC Triffid flax scare threatens access to no. 1 EU market. Manitoba Cooperator. http://www.manitobacooperator.ca/2009/09/17/cdc-triffid-flax-scare-threatens-access-to-no-1-eu-market/. Published September 17, 2009.
30. Dawson A. Changes likely for flax industry. Manitoba Cooperator. http://www.gmwatch.org/

component/content/article/11541. Published September 24, 2009.
31. Franz-Warkentin P. Flax industry sees "good progress" against Triffid. AGCanada. http://www.agcanada.com/daily/flax-industry-sees-good-progress-against-triffid-2. Published October 12, 2011.
32. Laidlaw S. StarLink fallout could cost billions: Future of modified crops thrown in doubt, report says. The Toronto Star. http://www.mindfully.org/GE/StarLink-Fallout-Cost-Billions.htm. Published January 9, 2001.
33. Centers for Disease Control and Prevention (CDC). Investigation of Human Health Effects Associated with Potential Exposure to Genetically Modified Corn: A Report to the US Food and Dug Administration; 2001. www.cdc.gov/nceh/ehhe/cry9creport/pdfs/cry9creport.pdf.
34. FIFRA Scientific Advisory Panel. A Set of Scientific Issues Being Considered by the Environmental Protection Agency Regarding Assessment of Additional Scientific Information Concerning StarLinkTM Corn. SAP Report No. 2001-09. Arlington, Virginia: US Environmental Protection Agency (EPA); 2001.
35. Carpenter JE, Gianessi LP. Agricultural Biotechnology: Updated Benefit Estimates. National Center for Food and Agricultural Policy; 2001. http://ucbiotech.org/biotech_info/PDFs/Carpenter_2001_Updated_Benefits.pdf.
36. Elsanhoty RM, Al-Turki AI, Ramadan MF. Prevalence of genetically modified rice, maize, and soy in Saudi food products. Appl Biochem Biotechnol. August 2013. doi:10.1007/s12010-013-0405-x.
37. Bondera M, Query M. Hawaiian Papaya: GMO Contaminated. Hawaii SEED; 2006. http://hawaiiseed.org/wp-content/uploads/2012/11/Papaya-Contamination-Report.pdf.
38. Strauss E. Genetically modified coffee confrontation brewing in Hawaii. PCC Nat Mark. June 2009. http://www.pccnaturalmarkets.com/sc/0906/sc0906-coffee-gm-hawaii.html.
39. Harmon A. A lonely quest for facts on genetically modified crops. The New York Times. http://www.nytimes.com/2014/01/05/us/on-hawaii-a-lonely-quest-for-facts-about-gmos.html. Published January 4, 2014.
40. Charles D. In Oregon, the GMO wheat mystery deepens. NPR.org. July 2013. http://www.npr.org/blogs/thesalt/2013/07/17/202684064/in-oregon-the-gmo-wheat-mystery-deepens.
41. Dupont V. Monsanto testing new GM wheat after 8-year freeze (Update). Phys.org. http://phys.org/news/2013-06-monsanto-gmo-wheat.html. Published June 5, 2013.
42. Monsanto. GM wheat questions and answers. 2014. http://www.monsanto.com/gmwheat/pages/gm-wheat-questions-and-answers.aspx#six.
43. US Department of Agriculture (USDA). USDA Announces Close and Findings of Investigation into the Detection of Genetically Engineered Wheat in Oregon in 2013. Riverdale, MD; 2014. http://www.aphis.usda.gov/newsroom/2014/09/pdf/ge_wheat.pdf.
44. Organic Agriculture Protection Fund Committee. Organic farmers seek Supreme Court hearing. August 2007. http://bit.ly/1iGdQla.
45. Binimelis R. Coexistence of plants and coexistence of farmers: Is an individual choice possible? J Agric Environ Ethics. 2008;21:437-457.
46. US Department of Agriculture Office of Inspector General Southwest Region. Audit Report: Animal and Plant Health Inspection Service Controls over Issuance of Genetically Engineered Organism Release Permits. Washington, DC: US Department of Agriculture Office of Inspector General Southwest Region; 2005. http://www.usda.gov/oig/webdocs/50601-08-TE.pdf.
47. Lambrecht B. Gene-altered apple tested in Washington state. Seattlepi.com. http://www.seattlepi.com/local/article/Gene-altered-apple-tested-in-Washington-state-5736742.php. Published September 5, 2014.
48. Lambrecht B. GMO experiments receive questionable oversight: Central Coast corn used for varied experiments. SFGate. http://m.sfgate.com/science/article/GMO-experiments-receive-questionable-oversight-5740478.php. Published September 7, 2014.

13 **Myth:** GM crops are needed to feed the world

Truth: GM crops are irrelevant to food security

Myth at a glance

GM crops are not needed to feed the world. We already produce enough food for 14 billion people, far more than we will ever need to feed the projected world population of 9 billion in 2050. People are hungry not because of a shortage of food production, but because of poverty: they cannot afford to buy food and lack the land on which to grow it.

Conventional plant breeding continues to outperform GM in producing crops with high yield and other useful traits, such as tolerance to extreme weather conditions and poor soils, complex-trait disease resistance, and enhanced nutritional value. These complex traits involve many genes working together in a precisely regulated way. It is difficult or impossible to successfully genetically engineer them into crops.

Several experiments involving GM crops targeted at poor and small-scale African farmers have ended in failure. Meanwhile non-GMO alternatives have been developed at a fraction of the cost and in a fraction of the time required to produce GMO versions. Many are already making a difference in farmers' fields.

The best known example of a GM crop with enhanced nutritional value is GMO golden rice, aimed at solving vitamin A deficiency. But after 15 years of broken promises and millions of dollars of investment, GMO golden rice is still not ready for farmers' fields. In 2014 it was announced that it had failed to give satisfactory yields in field trials. Meanwhile, non-GMO methods have successfully reduced vitamin A deficiency and only require modest funding to roll out more widely.

Everywhere we read that we must increase crop yields to feed the expanding world population. But this is a myth, a crisis narrative created by agribusiness interests in order to push the "solutions" of GM crops and chemically intensive farming.

There is no global or regional shortage of food – in India, South America, or Europe.[1] The world is swamped with food. According to experts advising the World Bank, we already produce enough food for 14 billion people, far more than we will ever need to feed the projected maximum world population of 9 billion in 2050.[2]

In the US, 40% of all food produced is wasted.[3] And the majority of the American GM crop harvest was never intended to feed people. Around 36% of the US maize crop goes into feed for US livestock and 40% into biofuels. Most of the rest is exported,[4] also for animal feed.[5] Only a tiny proportion goes to feed people, and most of that is in the form of high-fructose corn syrup,[4] an unhealthy ingredient of junk food.

Similarly, the vast acreages of GM soy and maize in Argentina and Brazil are used primarily for animal feed and industrial purposes. Animal feed does lead secondarily to human food, but with a massive loss of efficiency. Producing 1 kg of feedlot beef is estimated to require about 13 kg of grain.[6]

The cause of hunger is not a shortage of food production, but poverty. Even in countries where hunger is rife, there is plenty of food available in stores and markets for those who have money to buy it. Hunger is a social, political, and economic problem, which GM technology cannot address. In 2012 in India, millions of people went hungry while millions of tonnes of wheat and rice were left to rot in fields.[7] Claims that GM can solve hunger are a dangerous distraction from real solutions. Leaders in Africa have pointed to the biotech industry's claims that GM can solve the hunger problem as an exploitation of the suffering of the hungry.[8]

GM crops for Africa: Catalogue of failure

A handful of GM crops have been promoted as helping small-scale and poor farmers in Africa. However, the results were the opposite of what was promised.

GM sweet potato yielded poorly, lost virus resistance

The GM virus-resistant sweet potato was a GM showcase project for Africa. Florence Wambugu, the Monsanto-trained scientist fronting the project, was proclaimed the saviour of millions, based on her claims that the GM sweet potato had double the yield of non-GM sweet potatoes. Forbes

"[GMOs] haven't actually proven anything yet in terms of increased yields, as far as any of the major food crops are concerned... I don't really see any proper use for GMOs, now or even in the future... We produce enough food for 14 billion people."

– Dr Hans Herren, president of the Millennium Institute and co-chair, International Assessment of Agricultural Knowledge, Science and Technology, (IAASTD), a UN-, World Bank-, and WHO-sponsored project on the future of farming involving more than 400 experts from across the world[9]

"'Feeding the world' might as well be a marketing slogan for Big Ag, a euphemism for 'Let's ramp up sales,' as if producing more cars would guarantee that everyone had one... There are hungry people not because food is lacking, but because not all of those calories go to feed humans (a third go to feed animals, nearly 5% are used to produce biofuels, and as much as a third is wasted, all along the food chain)."

– Mark Bittman, food writer for the New York Times[10]

magazine declared her one of a handful of people around the globe who would "reinvent the future".[11]

Eventually it emerged that the GM sweet potato had failed its field trials. The GM sweet potato was out-yielded by the non-GM control and succumbed to the virus it was designed to resist.[12,13]

In contrast, a conventional breeding programme in Uganda produced a non-GM variety that was virus-resistant and raised yields by 100%. The Ugandan project achieved its goal in a fraction of the time – and at a fraction of the cost – of the GM project. The GM sweet potato project, over 12 years, consumed funding from Monsanto, the World Bank, and USAID to the tune of $6 million.[14]

GM cassava lost virus resistance

Cassava is a staple food in Africa. The potential of GM to boost cassava production by defeating a virus has been promoted since the 1990s. It was claimed that GM cassava could solve hunger in Africa by increasing yields as much as tenfold.[15] Even after it became clear that the GM cassava had suffered a technical failure and lost resistance to the virus,[16] media stories continued to appear about its curing hunger in Africa.[17,18]

Meanwhile, conventional (non-GM) plant breeding has quietly produced a virus-resistant cassava that is already proving successful in farmers' fields, even under drought conditions.[19]

Bt cotton failed in Makhatini

> *"The [GM cotton] seed itself is doing poorly. Without irrigation, and with increasingly unpredictable rain, it has been impossible to plant the cotton. In 2005 T.J. Buthelezi, the man whose progress was hymned by Monsanto's vice-president not three years before, had this to say: 'My head is full – I don't know what I'm going to do. I haven't planted a single seed this season. I have paid Rand 6,000 (USD 820, GBP 420) for ploughing, and I'm now in deep debt.' T.J. is one of the faces trucked around the world by Monsanto to prove that African farmers are benefiting from GM technology."*
>
> – Raj Patel, "Making up Makhatini", in *Stuffed and Starved*[20]

Makhatini in South Africa was home to a showcase GM Bt cotton project for small-scale farmers. The government-subsidized project began in 1997. By 2001 around 3,000 smallholder farmers cultivated Monsanto's Bt cotton.[21] The high rate of adoption was influenced by the fact that the only source of credit available to farmers was a company that was also the only buyer and seller of cotton and that promoted Bt cotton.[14]

The project failed due to adverse weather and farmer indebtedness.[14] Yields varied with rainfall, hovering within 10% of what they were before Bt cotton was introduced.[22] Pest attacks forced farmers to buy costly insecticide sprays.[23] Overall pest control costs were higher with Bt cotton (65% of total input costs) than with non-Bt cotton (42% of total input costs).[22] A 2003 report calculated that crop failures left the farmers who had adopted the expensive Bt cotton with debts of $1.2 million.[14]

In 2004, only 700 farmers were still growing Bt cotton, an 80% drop from the original figure.[24] By 2010–2011 the number of farmers had shrunk to 200. The area planted to Bt cotton had shrunk to 500 hectares – a 90% decline from the period of claimed success (1998–2000).[22]

A study published in 2006 concluded that the project did not generate a "tangible and sustainable socioeconomic improvement".[23] A 2012 review concluded that its main value appears to have been as a public relations exercise for GM proponents, providing "crucial ammunition to help convince other African nations to adopt GM crops".[22]

GM soy and maize project ends in ruin for poor farmers

A GM soy and maize farming project involving poor black farmers in South Africa was launched in 2003 with the support of the Eastern Cape government. But the project ended in disaster for the farmers, according to a study by the Masifunde Education and Development Project Trust and Rhodes University.[25] "We saw a deepening of poverty and people returning to the land for survival," said Masifunde researcher, Mercia Andrews.

The study raised concerns about feeding schemes conducted on animals with "alarming results", including damage to organs. It presented evidence of weed and pest problems and contamination of crops with GM pollen. It also drew attention to the problem of control over food systems exercised by big companies as a result of patented seeds.[25]

A peer-reviewed study published in 2015 examined the performance of Bt maize for smallholder farmers in South Africa. It found that "commercial varieties into which the Bt trait is introduced are outperformed by locally used non-GM hybrids and OPVs [open pollinated varieties], which are better adapted to smallholders' agroecologies, fluctuations in rainfall and suboptimal storage conditions."[26]

GM Bt brinjal fails two years running in Bangladesh

In 2014 the British pro-GMO campaigner Mark Lynas claimed that GM Bt brinjal (eggplant) was a success for poor farmers in Bangladesh in its first year of cultivation.[27] However, a local report from Bangladesh said the crop was a failure.[28]

The UK-based Guardian newspaper investigated, interviewing 19 out of the 20 farmers who grew GM Bt brinjal. The Guardian reported that while Bt brinjal successfully repelled the fruit and shoot borer pest as intended, some crop fields succumbed to bacterial wilt and drought. Nine of the 19 farmers said they had problems with the crop, with a failure rate of four out of five farms in one region.[29] The growing season culminated in a press conference in which furious farmers complained they had been "fooled" and used as guinea pigs for the failed crop, and demanded compensation for losses they said they had incurred.[30]

In 2015, in the second year of Bt brinjal cultivation, Lynas again claimed that the GM crop was a success. He wrote that a farmer growing Bt brinjal "had been able to stop using pesticides" thanks to the "new pest-resistant variety".[31]

But in this case too, local reports contradicted Lynas's claims. The Bangladesh daily New Age stated: "The cultivation of genetically engineered Bt brinjal in the country's several districts has cost the farmers their fortunes again this year as the plants have either died out prematurely or fruited very insignificantly compared to the locally available varieties."

The Bangladeshi NGO UBINIG also refuted Lynas's article, calling it "propaganda". UBINIG interviewed the farmer that Lynas had quoted and found that the Bt brinjals Lynas said were sold as "insecticide free" had actually been treated with pesticides. Moreover, the farmer, whom Lynas had claimed was being helped "out of poverty" by Bt brinjal, turned out to be a well-off commercial farmer. The farmer told UBINIG that he was "not happy" about the Bt brinjal's appearance, as it had a "rough surface and gets soft very quickly while the traditional variety is shiny and remains fresh for a longer time". The farmer added that 10% of the Bt brinjal plants were attacked by a virus.[32]

There are also toxicity questions over GM Bt brinjal. The developer company Mahyco's own tests revealed damage to the ovaries and spleen and toxic effects in the liver, as well as immune system changes, in Bt brinjal-fed rats. Dr Lou Gallagher, a New Zealand-based epidemiologist and risk assessment expert, commented that there were "major health problems among test animals... Release of Bt brinjal for human consumption cannot be recommended."[33]

Anna Lappé, founder of the Small Planet Institute, wrote that in contrast with the apparently failed Bt brinjal experiment, farmers using agroecological methods in Bangladesh and elsewhere are "achieving high yields with little to no use of chemical pesticides."[34]

GMO does not increase farmer choice but removes it

It is often claimed that farmers should be given the choice to grow GM crops. But farmers in countries that have adopted GM seeds have decreased seed choices, compared with those in countries that have remained free from GMO cultivation. In the countries where agbiotech companies gain entrance, they buy the main seed companies. The resulting consolidation in the seed market[35,36] has enabled the agbiotech companies to control which seed varieties farmers have access to – a process called choice editing. Usually this involves withdrawing high-performing non-GM seeds from the market so that farmers who want high-performance seeds must buy GM seeds. Companies may also set quotas. For instance, In Brazil, seed stores were required to sell 85% GMO soy seeds and no more than 15% non-GMO.[37] This trend of decreased seed choice with the arrival of GMO seed in a country has been documented in the US,[38] Brazil (with soy),[37] and India (with cotton).[39,40,41]

A study on farmer seed choice in Europe found that the GM-adopting country, Spain, had fewer seed choices on offer to farmers than non-GM adopting countries. Moreover, GM-adopting countries, including the US, had no yield advantage.[42]

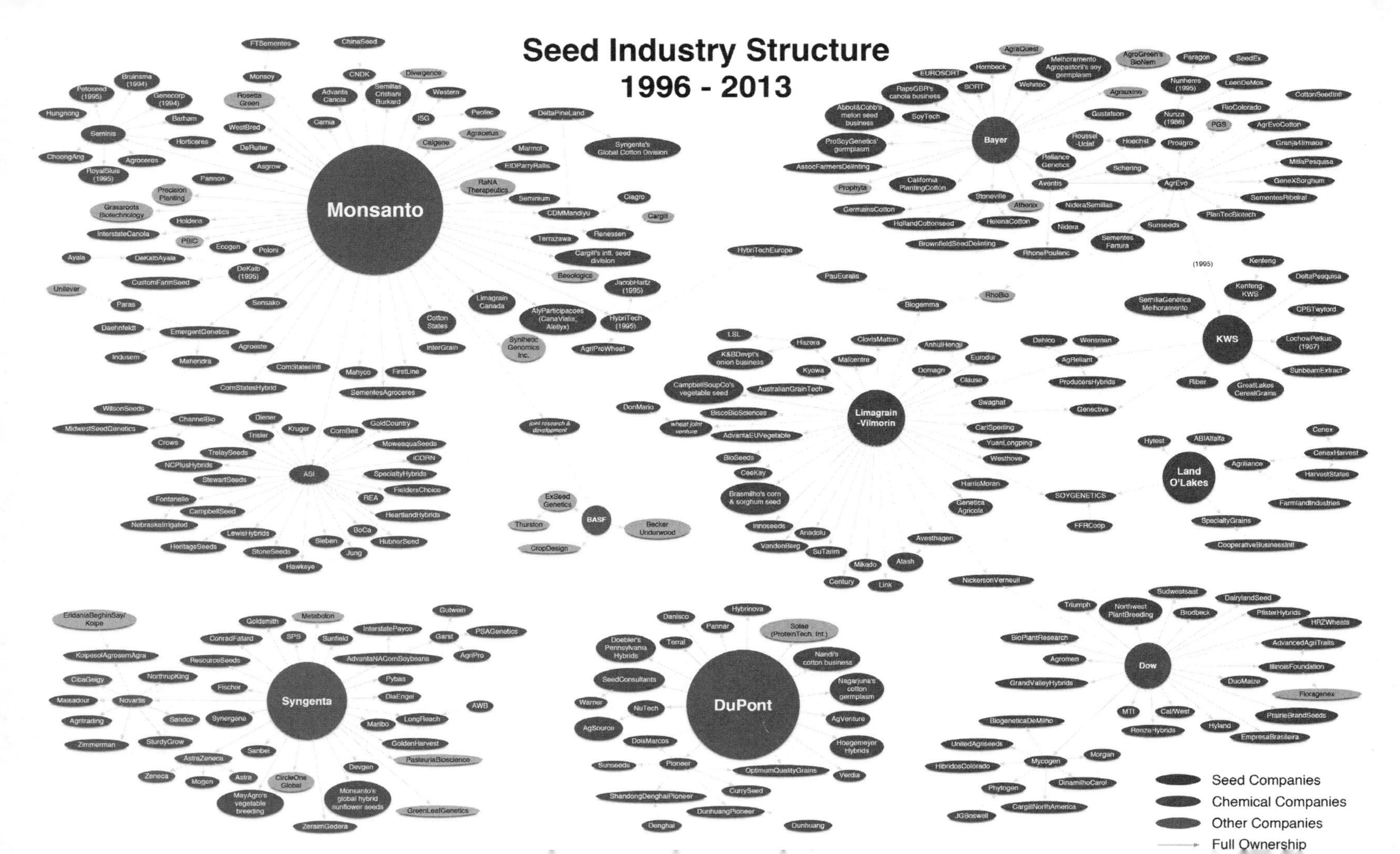

Seed Industry Structure
1996 - 2013
Monsanto
Bayer
KWS
Limagrain -Vilmorin
Land O'Lakes
ASI
BASF
Syngenta
DuPont
Dow
FTSementes
ChinaSeed
Monsoy
CNDK
Divergence
Bruinsma (1994)
Petoseed (1995)
Genecorp (1994)
Rosetta Green
Advanta Canola
Semillas Cristiani Burkard
Western
Hungnong
Barham
WestBred
Carnia
ISG
Peotec
Seminis
Horticeres
DeRuiter
Calgene
Agracetus
DeltaPineLand
ChoongAng
Agroceres
Asgrow
Marmot
Syngenta's Global Cotton Division
RoyalSluis (1995)
Pannon
EIDParryRallis
Precision Planting
RaNA Therapeutics
Seminium
Ciagro
Grassroots Biotechnology
Holdens
CDMMandiyu
Cargill
InterstateCanola
PBIC
Ecogen
Poloni
Terrazawa
Renessen
Ayala
DeKalbAyala
Cargill's intl. seed division
DeKalb (1995)
CustomFarmSeed
Beeologics
JacobHartz (1995)
Unilever
Paras
Sensako
Limagrain Canada
AlyParticipacoes (CanaVialis; Alellyx)
HybriTech (1995)
Cotton States
Daehnfeldt
EmergentGenetics
Synthetic Genomics Inc.
AgriProWheat
Agroeste
InterGrain
Indusem
Mahendra
CornStatesIntl
Mahyco
FirstLine
CornStatesHybrid
SementesAgroceres
WilsonSeeds
ChannelBio
Diener
Kruger
CornBelt
GoldCountry
MidwestSeedGenetics
Trisler
MoweaquaSeeds
Crows
TrelaySeeds
ICORN
NCPlusHybrids
SpecialtyHybrids
StewartSeeds
FieldersChoice
Fontanelle
REA
CampbellSeed
HeartlandHybrids
NebraskaIrrigated
LewisHybrids
BoCa
HeritageSeeds
Sieben
HubnerSeed
StoneSeeds
Jung
Hawkeye
joint research & development
ExSeed Genetics
Thurston
Becker Underwood
CropDesign
AgraQuest
Melhoramento Agropastoril's soy germplasm
AgroGreen's BioNem
Paragon
SeedEx
EUROSORT
Hornbeck
Nunhems (1995)
LeenDeMos
RapsGBR's canola business
SORT
Wehtec
Agrauxine
CottonSeedIntl
Abbot&Cobb's melon seed business
SoyTech
Gustafson
Nunza (1986)
RioColorado
PGS
AgrEvoCotton
ProSoyGenetics' germplasm
Roussel -Uclaf
Hoechst
Proagro
Granja4Irmaos
Reliance Genetics
AssocFarmersDelinting
Schering
MillaPesquisa
Prophyta
California PlantingCotton
Stoneville
Aventis
AgrEvo
GeneXSorghum
Athenix
SementesRibeiral
GermainsCotton
NideraSemillas
HollandCottonseed
HelenaCotton
PlanTecBiotech
Nidera
Sunseeds
BrownfieldSeedDelinting
Sementes Fartura
RhonePoulenc
HybriTechEurope
(1995)
Kenfeng
PauEuralis
DeltaPesquisa
Kenfeng-KWS
RhoBio
SemiliaGenética Melhoramento
Biogemma
CPBTwyford
LSL
ClovisMatton
Hazera
AnhuiHengji
Dahlco
Wensman
LochowPetkus (1967)
K&BDevpt's onion business
Malcentre
Eurodur
AgReliant
Kyowa
Domagri
SunbeamExtract
Clause
ProducersHybrids
Riber
GreatLakes CerealGrains
CampbellSoupCo's vegetable seed
AustralianGrainTech
DonMario
Swaghat
BiscoBioSciences
Genective
wheat joint venture
CarlSperling
AdvantaEUVegetable
YuanLongping
Hytest
ABIAlfalfa
Cenex
Westhove
BioSeeds
CenexHarvest
CeeKay
Agriliance
HarvestStates
HarrisMoran
Brasmilho's corn & sorghum seed
Genetica Agricola
SOYGENETICS
FarmlandIndustries
SpecialtyGrains
Innoseeds
FFRCoop
Anadolu
CooperativeBusinessIntl
VandenBerg
SuTarim
Avesthagen
Atash
Mikado
Century
Link
NickersonVerneuil
EridaniaBeghinSay/ Koipe
Metabolon
Gutwein
Goldsmith
InterstatePayco
Garst
PSAGenetics
ConradFafard
SPS
Sunfield
KoipesolAgrosemAgra
ResourceSeeds
AdvantaNACornSoybeans
AgriPro
CibaGeigy
NorthrupKing
Pybas
Fischer
DiaEngel
Maisadour
Novartis
AWB
Agritrading
Sandoz
Synergene
Maribo
LongReach
Zimmerman
SturdyGrow
GoldenHarvest
AstraZeneca
Sanbei
PasteuriaBioscience
Zeneca
Mogen
Astra
CircleOne Global
Devgen
MayAgro's vegetable breeding
Monsanto's global hybrid sunflower seeds
GreenLeafGenetics
ZeraimGedera
Hybrinova
Danisco
Pannar
Solae (ProteinTech. Int.)
Doebler's Pennsylvania Hybrids
Terral
Nandi's cotton business
SeedConsultants
Nagarjuna's cotton germplasm
Warner
NuTech
AgVenture
AgSource
DoisMarcos
Hoegemeyer Hybrids
Sunseeds
Pioneer
OptimumQualityGrains
Verdia
CurrySeed
ShandongDenghaiPioneer
DunhuangPioneer
Denghai
Dunhuang
Sudwestsaat
Triumph
Northwest PlantBreeding
DairylandSeed
Brodbeck
PfisterHybrids
HRZWheats
BioPlantResearch
AdvancedAgriTraits
Agromen
IllinoisFoundation
GrandValleyHybrids
DuoMaize
Floragenex
MTI
CalWest
BiogeneticaDeMilho
RenzeHybrids
Hyland
PrairieBrandSeeds
EmpresaBrasileira
UnitedAgriseeds
Morgan
HibridosColorado
Mycogen
Phytogen
DinamilhoCarol
JGBoswell
CargillNorthAmerica
Seed Companies
Chemical Companies
Other Companies
Full Ownership

Agroecology is the key to food security

Fortunately, there are plenty of alternatives to the GMO model of agriculture. In 2008 a ground-breaking four-year study on the future of farming was published. Sponsored by the World Bank and United Nations and conducted by over 400 international scientists, the International Assessment of Agricultural Knowledge, Science and Technology for Development (IAASTD) did not endorse GM crops as a solution to world hunger. The IAASTD report noted that yields of GM crops were "highly variable" and in some cases there were "yield declines". It added that safety questions remained over GM crops and that the patents attached to them could undermine seed saving and food security in developing countries.[43]

When asked at a press conference if GM crops were the answer to world hunger, IAASTD director Professor Bob Watson (subsequently chief scientist at the UK farm ministry Defra) said, "The simple answer is no."[44] The IAASTD concluded that the key to food security lies in agroecology.[43]

Agroecology includes low-input and organic methods that preserve soil and water and minimize external inputs. Scientific and technological advances are employed where appropriate.

Dramatic yield increases from agroecology

Agroecology projects in the Global South and other developing regions have produced dramatic increases in yields and food security.[45,46,47,48,49,50]

A 2008 United Nations report looked at 114 farming projects in 24 African countries and found that adoption of organic or near-organic practices resulted in yield increases averaging over 100%. In East Africa, a yield increase of 128% was found. The report concluded that organic agriculture can be more conducive to food security in Africa than chemically-based production systems, and that it is more likely to be sustainable in the long term.[48]

The System of Rice Intensification, known as SRI, is an agroecological

---

Chart (left): Consolidation has increased in the international seed industry in recent decades. The chart depicts changes in ownership involving major seed companies and their subsidiaries, primarily occurring from 1996 to 2013. The largest firms are represented as circles, with size proportional to global commercial seed market share.

Chart by Philip H. Howard, Associate Professor, Michigan State University. Chart reproduced with kind permission of Philip H. Howard. For an article providing more detail on earlier versions of this graphic, see: Howard, P. H. 2009. Visualizing consolidation in the global seed industry: 1996–2008. Sustainability 1(4), 1266-1287.

"Agroecology mimics nature not industrial processes. It replaces the external inputs like fertilizer with knowledge of how a combination of plants, trees and animals can enhance productivity of the land. Yields went up 214% in 44 projects in 20 countries in sub-Saharan Africa using agroecological farming techniques over a period of 3 to 10 years... far more than any GM crop has ever done."[53]

"Agroecological methods outperform the use of chemical fertilizers in boosting food production where the hungry live – especially in unfavorable environments. To date, agroecological projects have shown an average crop yield increase of 80% in 57 developing countries, with an average increase of 116% for all African projects. Recent projects conducted in 20 African countries demonstrated a doubling of crop yields over a period of 3–10 years. Conventional farming relies on expensive inputs, fuels climate change and is not resilient to climatic shocks. It simply is not the best choice any more. Agriculture should be fundamentally redirected towards modes of production that are more environmentally sustainable and socially just."[54]

– Olivier De Schutter, former UN special rapporteur on the right to food

method of increasing the productivity of irrigated rice by changing the management of plants, soil, water and nutrients. The benefits of SRI have been demonstrated in over 50 countries. They include yield increases of 20%–100%, up to a 90% reduction in required seed, and water savings of up to 50%.[51]

Wealthy countries too can benefit from innovative farming methods. In 2015 the South Australian agriculture minister reported the "amazing" results of a soil improvement programme, which he said resulted in yield increases of up to 300% for some crops.[52]

These results serve as a reminder that plant genetics are only a part of the answer to food security. The other part is how crops are grown. Agroecological farming methods not only ensure that there is enough food for the current population, but that the land improves in productivity to feed future generations.

Who owns food?

Traditionally, most of the seeds that farmers have used to feed humanity have not been owned by anyone. Farmers have been free to save seeds from one year's crop for the next year's crop, to share them with neighbours, and to produce them on a large scale to sell to others. Around 1.4 billion people in the Global

South still rely on farm-saved seed to feed their families and communities.[56]

But this ancient practice is being undermined, since the GM genes used in creating GM crops are patented and owned by GM companies. The patents forbid farmers from saving seeds containing the patented GM genes to plant the following year. They have to buy new seed each year.

Patents on GM seeds transfer control of food production from farmers to seed companies. GM companies co-opt centuries of farmer knowledge that went into creating locally adapted and genetically diverse seeds by adding one GM gene on top of the collective creation of generations of farmers.

Patents also transfer control of the food supply from the Global South to developed countries in the Global North, as most patents on food crops are owned by companies in developed countries in the North.[57] There is widespread concern in the Global South about the "biopiracy" of its genetic resources by the Global North, with the consequent loss of farmers' rights to save seed.

GM crops are not about feeding the world but about patented ownership of the food supply

While non-GM seeds are increasingly being patented, GM seeds opened the door to seed patenting, and are still easier to patent because the "inventive step" necessary to satisfy patent offices is clearer. From the beginning, patents on GM seeds have been a powerful lever that the agbiotech industry has used in consolidating its ownership of the seed industry and consequently its control of the food system.[58] According to the expert group ETC, just ten companies now control two-thirds of global seed sales.[59] These companies have the power to dictate which seeds are available to farmers to buy in some parts of the world today.

The patent granted on a GM gene sequence introduced into plant material extends to seeds, plants and any plants that are bred or otherwise derived (for example, by propagation) from the GM plant, all along the chain of farm and food production up to markets such as food and biofuels.[58] Patents make it possible to block access by other breeders to the genetic material. In comparison, the plant variety protection (PVP) system that has long applied to non-GM seeds allows others free access to commercial seed for further breeding.[58] Therefore the main objective of patents on GM plants is the monopolization of resources rather than the protection of inventions.[58]

In this context, the fact that GM fails to increase crop yields, reduce pesticide use, or deliver useful traits is of little concern to the companies that own the patents. As ETC said, "The new technologies don't need to be

"A key question for scientists and politicians to address is whether GM technology can co-exist in the global agricultural toolbox with other technologies without destroying those other tools. In 2002 the director of the Vietnamese government agricultural research centre told me that he could spend all of his annual R&D budget on lawyers, trying to sort out which materials his researchers could and could not use, and on licence fees for intellectual property rights (IPRs), according to the IPR jungle which has grown up around plant and crop materials. Is this kind of commercial restriction and narrowing of diversity of agricultural innovation trajectories helping food-poor countries to gain food security?"

– Professor Brian Wynne, Lancaster University, UK[55]

socially useful or technically superior (i.e. they don't have to work) in order to be profitable. All they have to do is chase away the competition and coerce governments into surrendering control. Once the market is monopolized, how the technology performs is irrelevant."[59]

The red herring of "public good" GMOs

Some GMO proponents say that the answer to patenting and ownership problems is for "public good" GM crops to be developed using public funds.[60] But this ignores the fact that through the mechanism of intellectual property (IP), public research institutions are converted into honorary corporations.

A public research institution developing a GMO licenses its patented technologies to the private sector to offset the cost of product development – the rest of the cost being paid by the taxpayer. So the taxpayer funding is given to chasing business-relevant research that never pays back to the public. And the public research institutions create internal incentives for academics to behave as salespeople for the GMO technology.

Beyond IP considerations, public institutions have come to rely on industry funding. GMO companies have representatives on university boards and fund research, buildings and departments.[61] This should be borne in mind when evaluating claims about GM technology by so-called independent scientists.

Whether a public institution or a private company develops a GMO has no effect on the research and development process or the outcome. Both types of entity will choose to develop IP on products that will make money rather than benefit farmers or the public. Both will use secrecy to protect the IP before it has been registered, thus screening out critical review by scientists and the public – who may not even want it. Both will charge for the product and for using it in breeding, to recover development costs and generate funds for other work.

Given that non-GMO solutions to the problems facing agriculture are readily available now, it seems irresponsible to gamble taxpayers' money on speculative GM "solutions".

Conventional breeding outperforms GM

The GMO lobby has been promising GM crops with desirable complex traits for 18 years. But today, almost all commercialized GMOs are engineered with one or both of two simple traits: to tolerate herbicides or to contain an insecticide.[62]

Conventional plant breeding continues to outperform GM in producing crops with useful traits such as high yields, tolerance to extreme weather conditions and poor soils, improved nutrient utilization, complex-trait disease resistance, and biofortification (enhanced nutritional value). Such traits are known as complex traits because they involve many genes working together in a precisely regulated way. Currently available genetic engineering methods are not precise or versatile enough to engineer the requisite large numbers of genes and regulatory networks into a plant.

Conventional breeding has achieved its successes more quickly and at a fraction of the cost of GM. In addition, GM is no quicker than conventional plant breeding and carries additional risks.

Drought-tolerant GM crops: No better than non-GM

Monsanto has released a drought-tolerant maize, but the US Department of Agriculture admitted that it was no more effective than existing non-GM varieties.[63] Syngenta's Agrisure Artesian drought-tolerant maize was developed using non-GM selective breeding from existing drought-tolerant non-GM maize varieties, but herbicide tolerant and insecticidal GM genes were subsequently added through genetic engineering.[64,65]

Truly drought-tolerant agriculture depends on agronomic methods more than genetics – for example, incorporating plenty of organic matter into the soil to enable it to absorb and retain water.

Non-GM virus-resistant papaya

A few disease-resistant GM crops have been developed, such as GM papayas engineered to resist the ringspot virus.[66,67] Resistance is conferred by inserting a single gene from the virus (virus coat protein gene) into the genome of the plant. When this protein is expressed in the papaya cells, it blocks the virus from infecting the cells.[68]

GMO proponents claim the GM papaya saved Hawaii's papaya industry.[69] But although GM papaya has dominated Hawaiian papaya production since the late 1990s, Hawaii's Department of Agriculture said the 2009 papaya production remained lower than when the ringspot virus was at its peak.[70] An article in the Hawaiian press said GM had not saved Hawaii's papaya industry, which had declined since 2002, possibly due to market rejection of GM foods.[71] In any case, non-GM ringspot virus-resistant papayas are available.[72]

In general, it is not necessary to use GM to develop virus-resistant crops, as conventional breeding has successfully developed such crops: for example, the virus-resistant sweet potato and cassava, discussed above.

The GM successes that never were

Many crops developed through conventional breeding, either alone or with assistance from marker assisted selection (MAS), are wrongly claimed as GM successes. MAS uses mapping of specific DNA sequences to track genetic markers (specific regions of DNA) associated with traits of interest during the conventional breeding process. This enables the breeder to quickly and precisely identify progeny that carry the genes and thus traits of interest.

Claimed "GM successes that never were" fall into three broad categories.

1. Conventionally bred crop with GM tweak

Typically, GM firms use conventional breeding, not GM, to develop crops with traits such as high yield or drought tolerance. By acquiring most of the leading seed companies, the agbiotech companies have obtained the germplasm of most of the best crop varieties developed over decades by seed breeders. They use conventional breeding and MAS to achieve the desired combination of complex traits. Once they have developed a successful variety using conventional breeding, they genetically modify that variety with company's proprietary GM genes. This GM tweak, usually a stack of herbicide-tolerance and insecticidal genes, adds nothing to the intrinsic yield of the crop. But it does allow the company to use the patents on the GM genes to prevent the variety from being produced or used for breeding by others.

As a spokesman for the Australian cereal breeding company, InterGrain, said: "Biotech traits by themselves are absolutely useless unless they can be put into the very best germplasm."[73]

An example of a GM product developed in this way is Monsanto's VISTIVE® soybean, which has been described as the first GM product with benefits for consumers. These low linolenic acid soybeans were developed by conventional breeding to produce oil that does not require the chemical treatments that generate harmful trans fats. It would have been a service to the public to offer this naturally bred soybean to farmers, but Monsanto did not do this. Instead it introduced a GM trait – tolerance to its Roundup herbicide[74] – and made only the GM variety available to farmers.

Iowa State University also used conventional breeding to develop soybean varieties with even lower levels of linolenic acid and made them available without GM traits.[75] However, since the university lacks the PR budget of Monsanto, very little has been heard about them, compared with VISTIVE.

2. Conventionally bred crop without GM tweak – GM used as lab tool

In some cases, a crop is developed with the aid of GM as a laboratory research tool to identify genes relevant to a certain trait. That gene is then used as a marker in Marker Assisted Breeding (MAS) to speed up the conventional breeding of a variety carrying the desired trait. No GM genes are added, but nevertheless, such crops have been claimed to be GM successes. An example is flood-tolerant rice, which the UK government's former chief scientist, Sir David King, wrongly claimed as a triumph of genetic engineering.[76,77]

The best-known flood-tolerant rice was developed by a research team led by Pamela Ronald.[78] Ronald's team developed the rice using MAS in a natural breeding programme.[78,79] They used genetic engineering to identify the desired genes and to look for their presence in the plants obtained through natural cross-breeding. The resulting rice is not genetically engineered but naturally bred.

Ronald does not appear to have tried to clear up the confusion – quite the opposite, as she misleadingly referred to the MAS method of developing the rice as "sort of a hybrid between genetic engineering and conventional breeding".[80] This is not the case. MAS is a tool that can be used with genetic engineering and conventional breeding, but it is not a "hybrid" of the two.

Another flood-tolerant rice, created with "Snorkel" genes, has also been claimed as a genetic engineering success.[81] But the rice, which adapts to flooding by growing long stems that prevent the crop from drowning, was bred by conventional methods and is non-GM.[82]

3. Crop that has nothing to do with GM

In one case, a crop that had nothing to do with GM at all was claimed as a GM success. The UK government's former chief scientist, Sir David King, wrongly claimed that a big increase in grain yields in Africa was due to GM.[83] In fact the yield increase was due to "push-pull", an agroecological method of companion planting that diverts pests away from crop plants.[84] King later admitted to what he called an "honest mistake".[85]

King produced this example when under pressure to provide reasons why GM crops are needed. In fact it shows the opposite – that we need to stop being distracted by GM and support effective non-GM solutions.

Non-GM breeding successes show no need for GM

The following are a few of many examples of conventionally bred crops with the types of traits that GMO proponents claim can only be achieved through genetic engineering. Many are already commercially available and making a difference in farmers' fields. The news and information service GMWatch keeps a more complete database of such non-GM breeding successes on its website.[86]

Drought-tolerant and climate-ready

- Maize varieties that yield well in drought conditions,[87] including some developed for Africa[88,89,90,91,92]
- Cassava that gives high yields in drought conditions and resists disease[19]
- Climate-adapted, high-yield sorghum varieties developed for Mali[93]
- Beans resistant to heat, drought, and disease[94,95]
- Pearl millet, sorghum, chickpea, pigeon pea and groundnut varieties that tolerate drought and high temperatures[96]
- Rice varieties bred to tolerate drought, flood, disease, and saline (salty) soils[97]
- Drought-resistant rice that yields up to 30% higher than other local varieties in Uganda[98]
- Flood-tolerant rice varieties developed for Asia[99,100]
- Over 2,000 indigenous rice varieties adapted to environmental fluctuations and resistant to pests and diseases, registered by Navdanya, an NGO based in India[101]
- Tomato varieties developed by Nepali farmers that tolerate heat and resist disease[102]
- Durum wheat that yields 25% more in saline soils than a commonly used variety.[103,104]

High-yield, pest-resistant, and disease-resistant

- High-yield, multi-disease-resistant beans for Central and East Africa[105]
- High-yield, disease-resistant cassava for Africa[106]
- Australian high-yield maize varieties targeted at non-GM Asian markets[107]
- Maize that resists the Striga parasitic weed pest and tolerates drought and low soil nitrogen, developed for Africa[90]
- Maize that resists the larger grain borer pest[108]
- "Green Super-Rice" bred for high yield and disease resistance[97]
- High-yield soybeans that resist the cyst nematode pest[109]
- Aphid-resistant soybeans[110,111,112,113]
- High-yield tomato with sweeter fruit[114]
- High-yield, pest-resistant chickpeas[115]
- Sweet potato that is resistant to nematodes, insect pests and Fusarium wilt, a fungal disease[116]
- High-yield, high-nutrition, and pest-resistant "superwheat"[117]
- Potato that resists root-knot nematodes[118]
- Potatoes that resist late blight and other diseases.[119,120,121,122,123,124,125]

Nutritionally fortified and health-promoting

- Soybeans with high levels of oleic acid, reducing the need for hydrogenation, which leads to the formation of unhealthy trans fats[131]
- Beta-carotene-enriched orange maize to combat vitamin A deficiency[132,133]
- Millet rich in iron, wheat abundant in zinc, and beta-carotene-enriched cassava[134]
- Purple potatoes containing high levels of the cancer-fighting antioxidants, anthocyanins[135,136]
- A tomato containing high levels of an antioxidant that may help combat heart attacks, stroke, and cancer[138]
- A purple tomato containing high levels of anthocyanins and vitamin C[137] (this attracted a fraction of the publicity gained by the GM purple

Rothamsted's GMO aphid-repellent wheat trial a £1 million failure

Genetics are only part of the solution to pest problems. Sometimes they are just a distraction from existing agronomic solutions. For example, Rothamsted Research in the UK has trialled wheat genetically engineered to produce a chemical that repels aphids. Rothamsted presented the project, which swallowed £1.28 million of public funds,[126] as an environmentally friendly way to control aphids.[127] But this was a GM "solution" to a problem that had already been solved by agroecology. Previous research by Rothamsted and others has shown that aphids in cereal crops can be successfully controlled by planting strips of flowers that attract natural predators.[128,129] In 2015 Rothamsted announced that its GMO project had failed: the GM wheat had not repelled aphids.[130]

"cancer-fighting" tomato[139,140,141])

- Low-allergy peanuts.[142] In a separate development, a process has been discovered to render ordinary peanuts allergen-free.[143]

What about GMO golden rice?

The most widely publicized example of a nutritionally fortified GM crop is golden rice, engineered to contain beta-carotene, which is converted to vitamin A once eaten. Golden rice is intended to help malnourished people who suffer from vitamin A deficiency. However, despite millions of dollars of investment and 15 years of broken promises of golden rice's imminent arrival, it has still not helped a single person, although it has been used to bolster the image of GM crops around the world.[144]

In May 2014 the International Rice Research Institute (IRRI), the body responsible for rolling out golden rice, announced that there would be a further delay in the timeline, due to golden rice having given disappointing yields in field trials. The IRRI admitted that the GMO rice had still not been safety tested, approved by regulators anywhere in the world, or shown to improve vitamin A status in malnourished people. All this has to be done before golden rice can be made available.[145]

Meanwhile, the Philippines, one of the countries targeted for golden rice, reduced vitamin A deficiency from 40% of children in 2003 to 15% in 2008 – just on the threshold of what is considered a public health problem.[146] This success was achieved using already available non-GMO methods such as vitamin supplements and food education.[147] The case of the Philippines shows that only modest funding and political will are needed to eradicate vitamin A deficiency across the Global South.

A representative of a Philippines farmer group that opposes GMO golden rice commented, "The Philippines is abundant with vitamin A-rich vegetables. In fact, these crops contain more beta-carotene than Golden Rice. So there's no reason for the government to commercialize this golden rice. We will not allow Filipinos to become guinea pigs of agrochemical TNCs [transnational companies]."[148]

Dr Michael Hansen, senior scientist with Consumers Union (USA), commented, "Vitamin A deficiency is a symptom of poverty – someone who is so poor they can only afford rice and virtually nothing else. Rather than treat the symptom, one should treat the cause – poverty." Hansen added that food diversification through sustainable agriculture and land reform are the long-term answers, since many local food crops have high beta-carotene levels,[147] yet malnourished people often lack access to land on which to grow them.

The main purpose of golden rice appears to be as a public relations strategy to break down resistance to GMOs.[149,150]

Consumer appeal

A GM non-browning apple has gained a less than enthusiastic reception from the apple industry, which fears it could damage markets.[151,152] Meanwhile, a non-GM version is available.[153]

GM is no quicker than conventional breeding – but it is more expensive

> *"The assertion that GM is quicker than breeding is common, but false. The average time required to develop a non-vegetatively engineered crop is about the same as developing one produced through breeding."*
>
> – Doug Gurian-Sherman, biotechnology specialist at the Union of Concerned Scientists[154]

> *"The overall cost to bring a new biotech trait to the market between 2008 and 2012 is on average $136 million."*
>
> – Phillips McDougall, in a consultancy study for the GMO industry[155]

A leading maize breeder, Major M. Goodman of North Carolina State University, says GM is no quicker than conventional breeding and that GM involves additional steps. He states that provided there are no complications, there is a 10–15-year lag time between the discovery of a new, potentially useful transgene and seed sales to farmers – about the same as the time needed to breed a new variety of a sexually propagated non-GM crop.[156,157] The Bt insecticidal trait engineered into GM crops took 16 years to reach the market – and that figure did not include the time required for toxicity testing.[156]

Dr Doug Gurian-Sherman of the Center for Food Safety agrees with this timescale: "When we look at actual examples, it has taken 10 to 15 years to develop a GM trait. And it is important to note that this is not an issue of delay due to regulatory requirements, as GM proponents are fond of asserting, but inherent in the limitations of the process."[154]

Gurian-Sherman explained, "Years of backcrossing are needed to get rid of possible harmful mutations and epigenetic changes introduced through the tissue culture process used with GM. And backcrossing is also needed to transfer the trait into elite crop varieties."[154]

In addition, years of field testing are needed to determine how well the crop responds in different environments, regardless of whether it is GM or non-GM.[153]

As for the cost of GM versus non-GM plant breeding, an industry consultancy study put the cost of developing a GM trait and bringing it to market at $136 million. Only $35 million goes towards regulatory costs, the rest being taken up by basic research and development.[155] Even Monsanto admits that non-GM plant breeding is "significantly cheaper" than GM.[158] According to Major M. Goodman, the cost of developing a single-gene GM trait is fifty times as much as the cost of developing an equivalent conventionally bred plant variety. Goodman called the cost of GM breeding a "formidable barrier" to its expansion.[156]

References

1. Latham J. How the Great Food War will be won. Independent Science News. http://www.independentsciencenews.org/environment/how-the-great-food-war-will-be-won/. Published January 12, 2015.
2. World Bank Institute. WBI Global Dialogue on Adaptation and Food Security: Summary of Main Issues. Washington, DC: World Bank Institute; 2011. http://wbi.worldbank.org/wbi/Data/wbi/wbicms/files/drupal-acquia/wbi/WBI%20Global%20Dialogue%20on%20Food%20Security%20and%20Adaptation%20-%20Summary%20of%20emerging%20issues.pdf.
3. Hall KD, Guo J, Dore M, Chow CC. The progressive increase of food waste in America and its environmental impact. PLoS ONE. 2009;4(11):e7940. doi:10.1371/journal.pone.0007940.
4. Foley J. It's time to rethink America's corn system. Sci Am. March 2013. http://www.scientificamerican.com/article/time-to-rethink-corn/.
5. Carter CA, Miller HI. Corn for food, not fuel. The New York Times. http://www.nytimes.com/2012/07/31/opinion/corn-for-food-not-fuel.html. Published July 30, 2012.
6. Pimentel D, Pimentel M. Sustainability of meat-based and plant-based diets and the environment. Am J Clin Nutr. 2003;78(3):660S - 663S.
7. Bhardwaj M. As crops rot, millions go hungry in India. Reuters. http://uk.reuters.com/article/2012/07/01/uk-india-wheat-idUKBRE8600KB20120701. Published July 1, 2012.

8. Paul H, Steinbrecher R. Hungry Corporations: Transnational Biotech Companies Colonise the Food Chain. London, UK: Zed Books; 2003.
9. Sherman M. Q & A: Hans Herren on Sustainable Agriculture Solutions. GMO Inside. http://gmoinside.org/q-hans-herren-sustainable-agriculture-solutions/. Published April 9, 2014.
10. Bittman M. How to feed the world. New York Times. http://www.nytimes.com/2013/10/15/opinion/how-to-feed-the-world.html. Published October 14, 2013.
11. Cook LJ. Millions served. Forbes. December 2002. http://www.forbes.com/forbes/2002/1223/302.html.
12. Gathura G. GM technology fails local potatoes. The Daily Nation (Kenya). http://bit.ly/KPQPxL. Published January 29, 2004.
13. New Scientist. Monsanto failure. 2004;181(2433). http://bit.ly/MHPG9W.
14. deGrassi A. Genetically Modified Crops and Sustainable Poverty Alleviation in Sub-Saharan Africa: An Assessment of Current Evidence. Third World Network – Africa; 2003. http://allafrica.com/sustainable/resources/view/00010161.pdf.
15. Groves M. Plant researchers offer bumper crop of humanity. LA Times. http://articles.latimes.com/1997/dec/26/news/mn-2352. Published December 26, 1997.
16. Donald Danforth Plant Science Center. Danforth Center cassava viral resistance review update. June 2006. http://bit.ly/1ry2DUC.
17. Greenbaum K. Can biotech from St. Louis solve hunger in Africa? St. Louis Post-Dispatch. http://bit.ly/L2MmG4. Published December 9, 2006.
18. Hand E. St Louis team fights crop killer in Africa. St Louis Post-Dispatch. http://www.gmwatch.org/index.php/news/archive/2006/5580. Published December 10, 2006.
19. International Institute of Tropical Agriculture (IITA). Farmers get better yields from new drought-tolerant cassava. http://bit.ly/L3s946. Published November 3, 2008.
20. Patel R. Making up Makhatini. In: Stuffed and Starved. Vol London, UK: Portobello Books; 2007:153-158.
21. Falck-Zepeda J, Gruère G, Sithole-Niang I. Genetically modified crops in Africa: Economic and policy lessons from countries south of the Sahara. In: Vol International Food Policy Research Institute (IFPRI); 2013:27-29. http://www.ifpri.org/sites/default/files/publications/oc75.pdf.
22. Schnurr MA. Inventing Makhathini: Creating a prototype for the dissemination of genetically modified crops into Africa. Geoforum. 2012;43(4):784-792.
23. Hofs J-L, Fok M, Vaissayre M. Impact of Bt cotton adoption on pesticide use by smallholders: A 2-year survey in Makhatini Flats (South Africa). Crop Prot. 2006;25:984-988.
24. Pschorn-Strauss E. Bt Cotton in South Africa: The Case of the Makhathini Farmers. Durban, South Africa: Biowatch South Africa; 2005. http://www.grain.org/article/entries/492-bt-cotton-in-south-africa-the-case-of-the-makhathini-farmers.
25. Jack M. GM project faces ruin. The New Age (South Africa). http://www.thenewage.co.za/21688-1008-53-GM_project_faces_ruin. Published June 28, 2011.
26. Fischer K, van den Berg J, Mutengwa C. Is Bt maize effective in improving South African smallholder agriculture? South Afr J Sci. 2015;111(1/2):1-2. doi:10.17159/sajs.2015/a0092.
27. Lynas M. Bt brinjal in Bangladesh – the true story. marklynas.org. http://www.marklynas.org/2014/05/bt-brinjal-in-bangladesh-the-true-story/. Published May 8, 2014.
28. Maswood MH. Bt brinjal farming ruins Gazipur farmers. New Age (Bangladesh). http://newagebd.net/9116/bt-brinjal-farming-ruins-gazipur-farmers/#sthash.3lZLCkrO.dpbs. Published May 7, 2014.
29. Hammadi S. Bangladeshi farmers caught in row over $600,000 GM aubergine trial. The Guardian. http://www.theguardian.com/environment/2014/jun/05/gm-crop-bangladesh-bt-brinjal. Published June 5, 2014.
30. New Age. Bt brinjal farmers demand compensation. New Age. http://newagebd.net/44155/bt-brinjal-farmers-demand-compensation/#sthash.sbn4tRM7.QtjuGPLK.dpbs. Published September 1, 2014.
31. Lynas M. How I got converted to GMO food. The New York Times. http://www.nytimes.com/2015/04/25/opinion/sunday/how-i-got-converted-to-gmo-food.html. Published April 24, 2015.
32. Akhter F. Turning Bt brinjal failure into a propaganda of success. ubinig.org. http://ubinig.org/index.php/home/showAerticle/76/english. Published May 15, 2015.
33. Gallagher LM. Bt Brinjal Event EE1: The Scope and Adequacy of the GEAC Toxicological Risk Assessment. Wellington, New Zealand; 2010. http://www.testbiotech.org/sites/default/files/Report%20Gallagher_2011.pdf.
34. Lappé A. Letter to the editor. New York Times. http://www.nytimes.com/2015/05/04/opinion/when-food-is-genetically-modified.html. Published May 4, 2015.
35. Howard P. Visualizing consolidation in the global seed industry: 1996–2008. Sustainability. 2009;1:1266-1287.
36. Howard P. Seed industry structure 1996–2013. Philip H Howard Assoc Profr Mich State Univ. 2014. https://www.msu.edu/~howardp/seedindustry.html.

37. Patriat P. Speech delivered at the association of seed producers of Mato Grosso, on May 11, 2011 at the soy industry conference SEMEAR 2011 in Sao Paulo, Brazil. GMWatch. July 2012. http://www.gmwatch.org/latest-listing/1-news-items/14092.
38. Hart M. Farmer to farmer: The truth about GM crops [film]. http://gmcropsfarmertofarmer.com/film.html. Published 2011.
39. Roseboro K. Scientist: GM technology has exacerbated pesticide treadmill in India. The Organic & Non-GMO Report. http://www.non-gmoreport.com/articles/february2012/gmtechnologypesticideindia.php. Published February 1, 2010.
40. Aaronson T. The suicide belt. Columbia City Paper. http://www.gmfreecymru.org.uk/documents/suicidebelt.html. Published November 10, 2009.
41. Disappearing Non-GM Cotton – Ways Forward to Maintain Diversity, Increase Availability and Ensure Quality of Non-GM Cotton Seed. Karnataka, India: Research Institute of Organic Agriculture (FiBL); 2011. http://www.fibl.org/fileadmin/documents/en/news/2011/ProceedingNationalWorkshop_DisappearingNon-GMCotton.pdf.
42. Hilbeck A, Lebrecht T, Vogel R, Heinemann JA, Binimelis R. Farmer's choice of seeds in four EU countries under different levels of GM crop adoption. Environ Sci Eur. 2013;25(1):12. doi:10.1186/2190-4715-25-12.
43. International Assessment of Agricultural Knowledge, Science and Technology for Development (IAASTD). Agriculture at a Crossroads: Synthesis Report of the International Assessment of Agricultural Knowledge, Science and Technology for Development: A Synthesis of the Global and Sub-Global IAASTD Reports. Washington, DC, USA: Island Press; 2009. http://www.unep.org/dewa/agassessment/reports/IAASTD/EN/Agriculture%20at%20a%20Crossroads_Synthesis%20Report%20%28English%29.pdf.
44. Lean G. Exposed: The great GM crops myth. The Independent. http://www.independent.co.uk/environment/green-living/exposed-the-great-gm-crops-myth-812179.html. Published April 20, 2008.
45. Altieri MA. Applying agroecology to enhance the productivity of peasant farming systems in Latin America. Environ Dev Sustain. 1999;1:197-217.
46. Bunch R. More productivity with fewer external inputs: Central American case studies of agroecological development and their broader implications. Environ Dev Sustain. 1999;1:219-233.
47. Pretty J. Can sustainable agriculture feed Africa? New evidence on progress, processes and impacts. J Environ Dev Sustain. 1999;1:253-274. doi:10.1023/A:1010039224868.
48. Hine R, Pretty J, Twarog S. Organic Agriculture and Food Security in Africa. New York and Geneva: UNEP-UNCTAD Capacity-Building Task Force on Trade, Environment and Development; 2008. http://bit.ly/KBCgY0.
49. Barzman M, Das L. Ecologising rice-based systems in Bangladesh. LEISA Mag. 2000;16. http://bit.ly/L2N71R.
50. Zhu Y, Chen H, Fan J, et al. Genetic diversity and disease control in rice. Nature. 17;406:718-722.
51. SRI International Network and Resources Center (SRI-Rice)/Cornell Univesity College of Agriculture and Life Sciences. Home page. 2014. http://sri.ciifad.cornell.edu/.
52. Grindlay D. SA agriculture minister says soil program proves utilising "God's gifts" can boost yields better than GM technology. ABC Rural. http://www.abc.net.au/news/2015-03-17/genetic-modification-grain-canola-agriculture-minister-bignell/6325276. Published March 17, 2015.
53. Leahy S. Africa: Save climate and double food production with eco-farming. IPS News. http://allafrica.com/stories/201103090055.html. Published March 8, 2011.
54. United Nations Human Rights Council. Eco-farming can double food production in 10 years, says new UN report [press release]. http://bit.ly/Lkfa9U. Published March 8, 2011.
55. Wynne B. Comment to Hickman, L., "Should the UK now embrace GM food?." The Guardian (UK). http://bit.ly/zvNSpL. Published March 9, 2012.
56. United Nations Development Programme (UNDP). Human Development Report 1999. New York and Oxford; 1999. http://hdr.undp.org/sites/default/files/reports/260/hdr_1999_en_nostats.pdf.
57. Khor M. In: Intellectual Property, Biodiversity, and Sustainable Development. Vol London, UK and Penang, Malaysia: Zed Books and Third World Network; 2002:9; 89.
58. Tippe R, Then C. Patents on melon, broccoli and ham? ELNI Rev. 2011;2:53-57.
59. ETC Group. Who Owns Nature? Corporate Power and the Final Frontier in the Commodification of Life. Ottawa, Canada; 2008. http://www.etcgroup.org/sites/www.etcgroup.org/files/publication/707/01/etc_won_report_final_color.pdf.
60. Jones JD. The cost of spurning GM crops is too high. The Guardian (UK). http://bit.ly/MpSIil. Published July 21, 2011.
61. Food & Water Watch. Public Research, Private Gain: Corporate Influence over University Agricultural Research. Washington, DC: Food & Water Watch; 2012. http://documents.

foodandwaterwatch.org/doc/PublicResearchPrivateGain.pdf.
62. James C. Global Status of Commercialized biotech/GM Crops: 2012. Ithaca, NY: ISAAA; 2012. http://www.isaaa.org/resources/publications/briefs/44/download/isaaa-brief-44-2012.pdf.
63. Voosen P. USDA looks to approve Monsanto's drought-tolerant corn. New York Times. http://nyti.ms/mQtCnq. Published May 11, 2011.
64. Ranii D. Drought-tough corn seed races to the finish line. newsobserver.com. http://bit.ly/KqA1xl. Published December 21, 2010.
65. Edmeades GO. Progress in Achieving and Delivering Drought Tolerance in Maize – An Update. Ithaca, NY: International Service for the Acquisition of Agri-biotech Applications (ISAAA); 2013. https://isaaa.org/resources/publications/briefs/44/specialfeature/Progress%20in%20Achieving%20and%20Delivering%20Drought%20Tolerance%20in%20Maize.pdf.
66. Gonsalves D. Transgenic papaya in Hawaii and beyond. AgBioForum. 2004;7(1 & 2):36-40.
67. Ferreira SA, Pitz KY, Manshardt R, Zee F, Fitch M, Gonsalves D. Virus coat protein transgenic papaya provides practical control of papaya ringspot virus in Hawaii. Plant Dis. 2002;86(2):101-105. doi:10.1094/PDIS.2002.86.2.101.
68. Fitch MMM, Manshardt RM, Gonsalves D, Slightom JL, Sanford JC. Virus resistant papaya plants derived from tissues bombarded with the coat protein gene of papaya ringspot virus. Nat Biotechnol. 1992;10(11):1466-1472. doi:10.1038/nbt1192-1466.
69. Summers J. GM halo effect: Can GM crops protect conventional and organic farming? Genetic Literacy Project. http://www.geneticliteracyproject.org/2014/01/09/gm-papaya-halo-effect/#.U2Kp3ccowsk. Published January 9, 2014.
70. Chan K. War of the papayas. ChinaDaily.com. http://bit.ly/LQT67d. Published September 8, 2011.
71. Hao S. Papaya production taking a tumble. The Honolulu Advertiser. http://bit.ly/LzDZRb. Published March 19, 2006.
72. Siar SV, Beligan GA, Sajise AJC, Villegas VN, Drew RA. Papaya ringspot virus resistance in Carica papaya via introgression from Vasconcellea quercifolia. Euphytica. 2011;181(2):159-168.
73. ABC Rural News Online. Monsanto and the WA Government Team up on Grain Breeding: Skye Shannon Speaks with Brian Whan, Intergrain and Peter O'Keefe, Monsanto [Audio].; 2010.
74. PR Newswire. Cargill to process Monsanto's VISTIVE(TM) low linolenic soybeans. http://prn.to/KyIREy. Published October 4, 2005.
75. Iowa State University. Six new soybean varieties highlight progress in developing healthier oils at ISU. http://www.plantbreeding.iastate.edu/pdf/soybeanReleases11-08.pdf. Published 2008.
76. Melchett P. Who can we trust on GM food? The Guardian (UK). http://www.guardian.co.uk/commentisfree/2008/dec/09/david-king-gm-crops. Published December 9, 2008.
77. Pendrous R. Europe's GM barrier is "starving the poor." FoodManufacture.co.uk. http://bit.ly/MpPw6m. Published June 13, 2011.
78. Xu K, Xu X, Fukao T, et al. Sub1A is an ethylene-response-factor-like gene that confers submergence tolerance to rice. Nature. 2006;442:705-708. doi:10.1038/nature04920.
79. Gunther M. Biotech and organic food: A love story. marcgunther.com. http://www.marcgunther.com/biotech-and-organic-food-a-love-story/. Published April 15, 2010.
80. Lebwohl B. Pamela Ronald has developed a more flood-tolerant rice. EarthSky. July 2010. http://earthsky.org/food/pamela-ronald-has-developed-a-more-flood-tolerant-rice.
81. Hsu J. Genetically engineered rice plants grow "snorkels" to survive floods. Pop Sci. August 2009. http://www.popsci.com/scitech/article/2009-08/snorkel-genes-allow-rice-plants-survive-floods.
82. Hattori Y, Nagai K, Furukawa S, et al. The ethylene response factors SNORKEL1 and SNORKEL2 allow rice to adapt to deep water. Nature. 2009;460:1026-1030.
83. BBC Today Programme. David King interviewed by Sarah Montague. November 2007.
84. Rothamsted Research Chemical Ecology Group. Push-pull habitat manipulation for control of maize stemborers and the witchweed Striga. http://bit.ly/1pST9I5.
85. Adam D. Eco Soundings: It's in the Mail. The Guardian (UK). http://www.guardian.co.uk/environment/2008/jul/30/1. Published July 30, 2008.
86. GMWatch. Non-GM successes. 2014. http://www.gmwatch.org/index.php/articles/non-gm-successes.
87. Gillam C. DuPont says new corn seed yields better in droughts. Reuters. http://reut.rs/Li0c5B. Published January 5, 2011.
88. Cocks T. Drought tolerant maize to hugely benefit Africa: Study. Reuters Africa. http://bit.ly/bPXW0p. Published August 26, 2010.
89. La Rovere R, Kostandini G, Tahirou A, et al. Potential Impact of Investments in Drought Tolerant Maize in Africa. Addis Ababa, Ethiopia: CIMMYT; 2010. http://bit.ly/1mLExYQ.
90. Atser G. Ghanaian farmers get quality protein, drought-tolerant, and Striga-resistant maize varieties to boost production. Modern Ghana. http://bit.ly/LZolNL. Published April 2, 2010.

91. Khisa I. Drought tolerant maize varieties ready. The East African. http://www.www.theeastafrican.co.ke/news/Drought-tolerant-maize-varieties-ready/-/2558/2134334/-/yk6a9p/-/index.html. Published January 4, 2014.
92. Atser G. Nigeria releases two extra-early maturing white maize hybrids. modernghana.com. http://www.modernghana.com/news/482841/1/nigeria-releases-two-extra-early-maturing-white-ma.html. Published August 17, 2013.
93. Diarra ST. Resistant seed helps Mali farmers battling climate change. AlertNet. http://bit.ly/Li0AkE. Published January 11, 2011.
94. Yao S. ARS releases heat-tolerant beans. USDA Agricultural Research Service. http://www.ars.usda.gov/is/pr/2010/100630.htm. Published June 30, 2010.
95. USDA Agricultural Research Service. Help for the common bean: Genetic solutions for legume problems. Agric Res USDA. 2010;May-June. http://www.ars.usda.gov/is/ar/archive/may10/bean0510.htm.
96. International Crops Research Institute for the Semi-Arid Tropics (ICRISAT). ICRISAT develops climate change ready varieties of pearl millet, sorghum, chickpea, pigeonpea and groundnut. SeedQuest. http://bit.ly/KqvVoV. Published June 5, 2009.
97. Berthelsen J. A new rice revolution on the way? AsiaSentinel. http://bit.ly/Lzthdi. Published January 17, 2011.
98. Food and Agriculture Organization (FAO). Uganda: After decades of war, a new rice variety helps farmers resume their lives. http://www.fao.org/news/story/en/item/35606/icode/. Published October 2, 2009.
99. International Rice Research Institute (IRRI). Indian farmers adopt flood-tolerant rice at unprecedented rates. ScienceDaily. http://www.sciencedaily.com/releases/2010/09/100915151015.htm. Published September 15, 2010.
100. IRIN News. Philippines: Could flood-resistant rice be the way forward?http://www.irinnews.org/Report.aspx?ReportId=82760. Published February 5, 2009.
101. Commodity Online. GM and India's rice fields. http://www.rediff.com/money/2007/mar/02comod4.htm. Published March 2, 2007.
102. Giri A. Nepali farm develops disease, heat resistant tomato. OneIndiaNews. http://news.oneindia.in/2010/12/05/nepalifarm-develops-disease-resistant-tomatoes.html. Published December 5, 2010.
103. Sawahel W. Wheat variety thrives on saltier soils. SciDevNet. April 2010. http://www.scidev.net/en/news/wheat-variety-thrives-on-saltier-soils.html.
104. Dean T. Salt tolerant wheat could boost yields by 25%. LifeScientist. http://lifescientist.com.au/content/biotechnology/news/salt-tolerant-wheat-could-boost-yields-by-25--583063808. Published March 12, 2012.
105. Ogodo O. Beans climb to new heights in Rwanda. SciDevNet. February 2010. http://www.scidev.net/en/news/beans-climb-to-new-heights-in-rwanda.html.
106. AFP. "Rooting" out hunger in Africa – and making Darwin proud. Indep UK. September 2010. http://www.independent.co.uk/life-style/health-and-families/rooting-out-hunger-in-africa--and-making-darwin-proud-2076547.html.
107. Queensland Country Life. New maize hybrids to target niche Asian markets. http://bit.ly/LZr89P. Published April 5, 2011.
108. CIMMYT. Body blow to grain borer. CIMMYT E-News. 2007;14 May 2012. http://www.cimmyt.org/en/news-and-updates/item/body-blow-to-grain-borer.
109. Swoboda R. Cho[o]se high-yielding, SCN-resistant soybeans. Wallace's Farmer (Iowa, USA). http://bit.ly/1fCi7H2. Published November 7, 2007.
110. Diers B. Discovering soybean plants resistant to aphids and a new aphid. University of Illinois Extension. http://web.extension.illinois.edu/state/newsdetail.cfm?NewsID=15202. Published February 20, 2010.
111. Li Y, Hill CB, Carlson SR, Diers BW, Hartman GL. Soybean aphid resistance genes in the soybean cultivars Dowling and Jackson map to linkage group M. Mol Breed. 2007;19(1):25-34. doi:10.1007/s11032-006-9039-9.
112. Kim K-S, Hill CB, Hartman GL, Mian MAR, Diers BW. Discovery of soybean aphid biotypes. Crop Sci. 2008;48(3):923. doi:10.2135/cropsci2007.08.0447.
113. Hill CB, Kim K-S, Crull L, Diers BW, Hartman GL. Inheritance of resistance to the soybean aphid in soybean PI 200538. Crop Sci. 2009;49(4):1193. doi:10.2135/cropsci2008.09.0561.
114. Allen J. Single gene powers hybrid tomato plants. PlanetArk. http://www.planetark.com/enviro-news/item/57360. Published March 30, 2010.
115. Suszkiw J. Experimental chickpeas fend off caterpillar pest. USDA Agricultural Research Service News & Events. http://www.ars.usda.gov/is/pr/2009/090825.htm. Published August 25, 2009.
116. Clemson University. New not-so-sweet potato resists pests and disease. Bioscience Technology. http://bit.ly/LGHVlo. Published June 22, 2011.
117. Kloosterman K. Pest-resistant super wheat "Al Israeliano." greenprophet.com. http://www.

greenprophet.com/2010/08/israel-super-wheat/. Published August 17, 2010.
118. Suszkiw J. Scientists use old, new tools to develop pest-resistant potato. USDA Agricultural Research Service. http://www.ars.usda.gov/is/ar/archive/apr09/potato0409.htm. Published March 31, 2009.
119. Clarke A. Conventional potato varieties resist PCN and blight. Farmers Wkly. April 2014. http://www.fwi.co.uk/articles/09/04/2014/144089/conventional-potato-varieties-resist-pcn-and-blight.htm.
120. Potato Council (UK). Toluca. Br Potato Var Database. 2014. http://varieties.potato.org.uk/display_description.php?variety_name=Toluca.
121. Wragg S. Elm Farm 2010: Blight-resistant spuds could lower carbon levels. Farmers Weekly Interactive. http://bit.ly/LsRjb2. Published January 11, 2010.
122. Suszkiw J. ARS scientists seek blight-resistant spuds. USDA Agricultural Research Service. http://www.ars.usda.gov/is/pr/2010/100603.htm. Published June 3, 2010.
123. Shackford S. Cornell releases two new potato varieties, ideal for chips. Chronicle Online. http://www.news.cornell.edu/stories/Feb11/NewPotatoes.html. Published February 21, 2011.
124. Fowler A. Sárpo potatoes. The Guardian. http://www.theguardian.com/lifeandstyle/2012/jan/13/alys-fowler-sarpo-potatoes. Published January 13, 2012.
125. White S, Shaw D. The usefulness of late-blight resistant Sarpo cultivars – A case study. ISHS Acta Hortic. 2009;834. http://www.actahort.org/members/showpdf?booknrarnr=834_17.
126. GM Freeze. Rothamsted's GM wheat – "A step backwards for farming" [press release]. http://www.gmfreeze.org/news-releases/187/. Published March 29, 2012.
127. Rothamsted Research. Rothamsted GM wheat trial. 2014. http://www.rothamsted.ac.uk/our-science/rothamsted-gm-wheat-trial.
128. Powell W, A'Hara SA, Harling R, et al. Managing Biodiversity in Field Margins to Enhance Integrated Pest Control in Arable Crops ("3-D Farming" Project): Project Report No. 356 Part 1. Home-Grown Cereals Authority (HGCA); 2004. http://archive.hgca.com/document.aspx?fn=load&media_id=1496&publicationId=1820.
129. Hickman JM, Written SD. Use of Phelia tanacetifolia strips to enhance biological control of aphids by overfly larvae in cereal fields. J Econ Entomol. 1996;89(4):832-840.
130. Bruce TJA, Aradottir GI, Smart LE, et al. The first crop plant genetically engineered to release an insect pheromone for defence. Sci Rep. 2015;5. doi:10.1038/srep11183.
131. Suszkiw J. New soybeans bred for oil that's more heart-healthy. USDA Agricultural Research Service News & Events. http://www.ars.usda.gov/is/pr/2010/100916.htm. Published September 16, 2010.
132. Li S, Nugroho A, Rocheford T, White WS. Vitamin A equivalence of the β-carotene in β-carotene–biofortified maize porridge consumed by women? Am J Clin Nutr. 2010;92(5):1105-1112. doi:10.3945/ajcn.2010.29802.
133. HarvestPlus. Scientists find that "orange" maize is a good source of vitamin A. HarvestPlus.org. http://bit.ly/L2PxNV. Published September 7, 2010.
134. Anderson T. Biofortified crops ready for developing world debut. SciDev.Net. http://bit.ly/MAkMg7. Published November 17, 2010.
135. BBC News. "Healthy" purple potato goes on sale in UK supermarkets. http://www.bbc.co.uk/news/uk-scotland-11477327. Published October 6, 2010.
136. Watson J. Purple spud will put you in the pink. Scotland on Sunday. http://scotlandonsunday.scotsman.com/uk/Purple-spud-will-put-you.4841710.jp. Published January 3, 2009.
137. Knowles M. Italian producers unveil "supertomato." Fruitnet.com. http://bit.ly/1oLKL7t. Published July 5, 2010.
138. CBS News. Purple tomatoes may fight cancer, other diseases. http://archive.digtriad.com/news/health/article/202115/8/Purple-Tomatoes-May-Fight-Cancer-Other-Diseases. Published December 3, 2011.
139. John Innes Centre. Purple tomatoes may keep cancer at bay. http://bit.ly/NAwtZ6. Published October 26, 2008.
140. Martin C. How my purple tomato could save your life. Mail Online. http://bit.ly/10JsmlO. Published November 8, 2008.
141. Derbyshire D. Purple "super tomato" that can fight against cancer. Daily Mail. http://www.athena-flora.eu/florapress/4-Purple_Tomatoes_International_press_clip/UK/daily%20mail_UK.pdf. Published October 27, 2008.
142. Asian News International. Low-allergy peanuts on the anvil. OneIndiaNews. http://bit.ly/Li7xlV. Published June 8, 2010.
143. North Carolina A&T State University School of Agricultural and Environmental Sciences. N.C. A&T food scientist develops process for allergen-free peanuts. EurekAlert. http://bit.ly/LQVQBE. Published July 23, 2007.
144. Matthews J. Golden rice: Is it vaporware? GMWatch. March 2015. http://gmwatch.org/index.php/news/archive/2015-articles/16043.
145. International Rice Research Institute (IRRI). What is the status of the Golden Rice project coordinated by IRRI? IRRI. May 2014. http://irri.org/golden-rice/faqs/what-is-the-status-of-the-golden-rice-project-coordinated-by-irri.

146. Food and Nutrition Research Institute/Dept of Science and Technology (Philippines). 7th National Nutrition Survey: 2008: Biochemical Survey Component. Manila, Philippines; 2010. http://www.fnri.dost.gov.ph/images/stories/7thNNS/biochemical/biochemical_vad.pdf.
147. Hansen M. Golden rice myths. GMWatch. http://gmwatch.org/index.php/news/archive/2013/15023. Published August 28, 2013.
148. Masipag Mindanao (Philippines). Farmers oppose golden rice; challenge foreign lobbyists to a debate. GMWatch. March 2015. http://gmwatch.org/index.php/news/archive/2015-articles/15985.
149. Matthews J. Transgenic Wars. GMWatch. http://www.gmwatch.org/index.php/news/archive/2015. Published April 7, 2015.
150. Biothai (Thailand), CEDAC (Cambodia), DRCSC (India), GRAIN, MASIPAG (Philippines), PAN-Indonesia and UBINIG (Bangladesh). Grains of Delusion: Golden Rice Seen from the Ground. Barcelona, Spain: GRAIN; 2001. http://www.grain.org/article/entries/10-grains-of-delusion-golden-rice-seen-from-the-ground.
151. Pollack A. That fresh look, genetically buffed. New York Times. http://www.nytimes.com/2012/07/13/business/growers-fret-over-a-new-apple-that-wont-turn-brown.html?_r=2&smid=tw-nytimesdining&seid=auto&. Published July 12, 2012.
152. Charles D. This GMO apple won't brown. Will that sour the fruit's image? NPR.org. January 2014. http://www.npr.org/blogs/thesalt/2014/01/08/260782518/this-gmo-apple-wont-brown-will-that-sour-the-fruits-image.
153. The editors of The Organic & Non-GMO Report. Washington State University develops non-GMO, non-browning apple alternative. The Organic & Non-GMO Report. http://www.non-gmoreport.com/articles/february2014/wsu-develops-non-gmo-non-browning-apple-alternative.php. Published January 31, 2014.
154. GMWatch. Is GM quicker than conventional breeding? http://www.gmwatch.org/index.php/news/archive/2013-2/15227. Published December 23, 2013.
155. Phillips McDougall. The Cost and Time Involved in the Discovery, Development and Authorisation of a New Plant Biotechnology Derived Trait: A Consultancy Study for Crop Life International. Pathhead, Midlothian; 2011.
156. Goodman MM. New sources of germplasm: Lines, transgenes, and breeders. In: Martinez JM, ed. Memoria Congresso Nacional de Fitogenetica. Vol Univ Autonimo Agr Antonio Narro, Saltillo, Coah, Mexico; 2002:28-41. http://www.cropsci.ncsu.edu/maize/publications/NewSources.pdf.
157. Goodman MM, Carson ML. Reality vs. myth: Corn breeding, exotics, and genetic engineering. In: Proc. of the 55th Annual Corn & Sorghum Research Conference. Vol 55. Chicago, IL; 2000:149-172.
158. Lloyd T. Monsanto's new gambit: Fruits and veggies. Harvest Public Media. http://bit.ly/LQTNxp. Published April 8, 2011.

Conclusion

GM crops and foods have been consistently promoted as a way to produce higher yields with less inputs, reduce pesticide use, make farming easier and more profitable, produce more nutritious foods, and meet the challenges of climate change.

But the evidence that has emerged since their introduction in 1996 paints a very different picture. Scientific research and real-world farming experience shows that GM crops have not delivered on these promises. They have not increased yields or sustainably reduced toxic chemical inputs. In reality they have presented farmers with the new challenges of controlling herbicide-resistant superweeds and Bt toxin-resistant super-pests. GM crops are no less dependent on artificial fertilizers than any other chemically grown crop. They are not as safe to eat as conventionally bred crops. They provide no solution to the major challenges of our time: climate change, the energy crisis, and world hunger.

Why has GM failed to deliver?

The GM approach treats genes as isolated units of information with predictable outcomes. But this approach is flawed. Gene organization within the DNA of any organism is not random and gene function is a complex, interconnected, and coordinated network, consisting of layer upon layer of molecular systems.

GM is based on an outdated understanding of genetics and is destined to fail. It is beyond the ability of GM to deliver anything but the simplest of properties such as single-gene herbicide tolerance. GM is simply not up to the task of delivering safe, productive, and resilient food production systems.

Our modern understanding of genetics tells us that we need to take a holistic approach to crop development that preserves gene organization and regulation, rather than disrupting it, as GM does. The way to safely and effectively generate crops with complex desirable properties such as higher yield, drought tolerance, and disease resistance is through natural breeding, helped where useful by marker assisted selection.

Why do farmers plant GM crops?

GMO proponents often argue that if GMOs were as unimpressive and problematic as we suggest, farmers would not plant them.

The simple answer to this argument is that while some farmers do plant GM crops, the vast majority do not. Non-GM farming is by far the dominant model. Industry figures from 2013 show that 18 million farmers grow GM crops in 27 countries worldwide: that's less than 1% of the farming population. Around 92% of all GMOs are grown in just six countries, and these countries mainly grow just four GM crops: soy, maize, oilseed rape (canola) and cotton. Eighty-eight percent of arable land across the globe remains GM-free.[1]

What is more, in 2014, industry figures revealed that GM crop planting had fallen in industrialized countries for the first time since the technology was commercialized in 1996. Clive James, head of the industry group ISAAA, said the industry now sees the developing world as the target for GMO industry expansion.[2]

But as the evidence and case studies presented in this report make clear, it is irresponsible to use farmers in the developing world as guinea pigs for experimental GM crops that the majority of people do not want to eat.

Time to move on

For two decades, GMO proponents have dominated the political and media discussion on food and agriculture. Many of our agricultural research institutes and universities accept GMO industry funding and obligingly pursue a narrow GM-focused agenda, at the expense of proven effective agroecological solutions that focus on improving soil quality and maintaining crop diversity and health.

Yet the public, the vast majority of whom do not want to eat GM foods, is unconvinced. It has become common for pro-GMO lobbyists to try to shut down resistance to GM food and agriculture by saying that the debate is over, that science has shown that GMOs are safe and beneficial, and that it is time to move on and accept them.

We agree with only one aspect of this argument. It is indeed time to move on, but in the opposite direction to the one promoted by the GMO proponents. The scientific evidence presented in this report shows that the hypothetical benefits of GM crops and foods are not worth the known risks.

It is time to face up to what the evidence tells us about GMOs and stop pretending that GMOs can do anything that non-GM agriculture and good farming can't do far better, at a fraction of the cost, and without the restrictions attached to patent ownership. In fact, patents represent the single area in which GM crops and foods outstrip non-GM. If it ever becomes as easy to patent a non-GM crop as it is to patent a GM crop, it is likely that

GM crops and foods will be consigned to the dustbin of history. Agricultural genetic engineering is not a smart or useful enough technology to succeed on its own merits. It is of interest to multinational companies and their government allies as a route to patented ownership of the food supply.

Once this fact has become clear to citizens and policymakers, we hope they will throw resources and funding behind the safe, sustainable, and equitable agriculture that the world needs.

References

1. Friends of the Earth. Who Benefits from GM Crops? An Industry Built on Myths. Amsterdam, The Netherlands; 2014. http://www.foeeurope.org/sites/default/files/publications/foei_who_benefits_from_gm_crops_2014.pdf.
2. Kaskey J. Modified crop plantings fall in industrialized nations. Bloomberg. http://www.bloomberg.com/news/2014-02-13/modified-crop-plantings-fall-in-industrialiez-nations.html. Published February 13, 2014.
3. Friends of the Earth. Who Benefits from GM Crops? An Industry Built on Myths. Amsterdam, The Netherlands; 2014. http://www.foeeurope.org/sites/default/files/publications/foei_who_benefits_from_gm_crops_2014.pdf.
4. Kaskey J. Modified crop plantings fall in industrialized nations. Bloomberg. http://www.bloomberg.com/news/2014-02-13/modified-crop-plantings-fall-in-industrialiez-nations.html. Published February 13, 2014.

Index